Sedimentary Crisis at the Global Scale 2

To Geneviève, for her support

Series Editor
Yves Lagabrielle

Sedimentary Crisis at the Global Scale 2

Deltas, a Major Environmental Crisis

Jean-Paul Bravard

iSTE

WILEY

First published 2019 in Great Britain and the United States by ISTE Ltd and John Wiley & Sons, Inc.

ISTE Ltd
27–37 St George's Road
London SW19 4EU
UK

www.iste.co.uk

John Wiley & Sons, Inc.
111 River Street
Hoboken, NJ 07030
USA

www.wiley.com

Library of Congress Control Number: 2018967007

British Library Cataloguing-in-Publication Data
A CIP record for this book is available from the British Library
ISBN 978-1-78630-384-4

FSC www.fsc.org
MIX
Paper from responsible sources
FSC® C013604

Contents

Introduction

Deltas are a major component of the Earth's "coastal fringe". They are sites where particulate and dissolved materials are deposited, and places where essential biochemical processes take place; their fragile and remarkable equilibrium arises from the complex set of constructive and destructive forces that have been at play at the interface between land and sea for a short period of time in geological terms, six millennia at the most. This gives rise to specific forms and varied typologies, depending on the relative importance of the influencing factors. Over the same period of time, watersheds and the rivers that they feed have been profoundly disrupted by deforestation, agriculture, extraction activities and fluvial engineering, often to the point of no longer being able to guarantee a sufficient contribution to balanced delta functionality, and according to the indications of all evidence, this trend will probably get worse. Finally, climate change is responsible for the rise in sea levels which increasingly attack delta environments that are then incapable of adjusting to the intensity of the process.

The second volume of "Sedimentary Crisis at the Global Scale" aims to examine the theme of exits to fluvial systems, in other words through deltas. As receptacles for sediments carried down from the continents, the Earth's deltas are very young zones when placed in the context of the scale of the Earth's geological history. However, they have a complex history in the era of continental sedimentary crisis, the reality of which we discussed in the first volume; deltas are at the mercy of the rhythm of rivers.

The crisis that the majority of the Earth's deltas are undergoing to a greater or lesser degree is in the continuity between places on the continent where flows of water, materials and dissolved substances are created, and the oceans that are natural receptacles for them. Societies have disrupted this continuity because, fundamentally, they have not understood it, or at least have not taken it into consideration.

These natural characteristics can come into conflict with the need for stability in contemporary human societies seeking to apply solutions to manage fluvial flooding and the actions of the sea. From an economic and human point of view, deltas and the peripheral region that they organize only cover 1% of the Earth's surface, but the figure for the population living there diverges; it would be between 10% and 25% of the global population. This is a way of illustrating their great importance for the planet, even if certain deltas in cold regions have been spared from settlement by humans. How can we explain why they are so attractive? Some possibilities are the fertility of the silts carried down by the annual floods, the ruling power of ancient civilizations or the drive of contemporary nations, and the ingenuity of traditional techniques for managing hydraulics and agricultural practices. Most deltas were then developed in such a way as to focus their economy on production of resources from underground, which poses the problems of compatibility with their internal equilibrium.

This second volume begins with *the physical operation of deltas*. The book briefly presents how deltas are formed, their structure and how the dynamics of water and materials flows are played out there. Their history, brief on a geological timescale, attests not only to their fragility but also to their instability right from the beginning of their construction, about 6,000 years ago. Presentation of three deltas of very different sizes and with unique watersheds, the Nile, the Huang-He (Yellow River) and the Rhône, will showcase a small portion of the diversity of situations that exist. Today, most deltas have been all but abandoned by continents since our societies have forgotten that they cannot survive without water and sediments, and they are threatened by rising sea levels and now more aggressive ocean behavior, such as the swell, waves and storms. Before examining deltas populated by humans, we will use satellite photographs to take a look at two deltas established in the Arctic, the Mackenzie Delta in Canada and the Lena Delta in Russia (Siberia). The climate still protects their basin and the area they cover from any strong influence on their environment. The last part of this chapter refers to a selection of situations of a radically different nature, selected from Southeast Asia. These are deltas occupied by large cities, where excessive sampling of water resources is responsible for their increased rate of subsidence*, in other words sinking that constantly reinforces the threat posed by the ocean.

Chapter 2 presents the cases of formerly occupied deltas in which crises have occurred during the Holocene* under the control of climatic factors (the Nile) and anthropological factors; the latter occurs at the scale of the watershed due to the variability of liquid and solid flows (the Huang-He and the Rhône, so different, but size matters little in the way the complexity is expressed). On a smaller timescale, more specifically in the most recent few millennia, the combined delta of the Rhine and the Meuse rivers displays a remarkable adaptation of human societies to both fluvial and marine constraints; at the scale of the last few years, during which the

Little Ice Age* has been experienced, the capacity of technicians and managers to manage the hydric and sedimentary constraints has been demonstrated as a function of the variation in design preferences through time. Other deltas around the world will be examined from the point of view of society's adaptation to fluvial overflow events under anthropological control; we have chosen the Nile after the construction of the Aswan Dam, the Rhône after the construction of multiple structures and reforestation of the mountains, the Ebro Delta deprived of sediments, the Po Delta with multiple weak points and finally the Danube Delta where a glimmer of hope remains.

Chapter 3 aims to instill an understanding of *the complex situation of deltas in the tropical regions*, sandwiched between an evolving continent and an ocean into which they open out. How can we manage the immense, highly populated areas of the Ganges and Brahmaputra deltas, exposed to cyclones and marine submersions while the rivers, either overused (the Ganges) or moving in that direction (the Brahmaputra), no longer guarantee the sedimentary input that is required? How can we manage the Mekong Delta, a victim of hydraulic developments in its basin which modify the hydrological and sedimentary regimes, even though management of the delta leaves a difficult legacy? The Mekong Delta is preparing in its own way to suffer a heavy impact from upstream and to adapt to the rise in sea level. The situation is even more dramatic in the Niger Delta, in the grip of a major political crisis due to the exploitation of oil resources, whether it be in the Indus Delta, deprived of water by irrigation in its basin, or even in the Ayeyarwady Delta, threatened by very uncertain developments. The stakes at hand affect tens of millions of people of very low income, living in nations that are either powerless or constrained by their neighbors upstream.

Chapter 4 is dedicated solely to a large delta, *the Mississippi*, since it has been studied widely and is therefore very well known, and above all because it is the epitome of an old delta with colonial roots, integrated into the most developed country in the world. Its modern history will be discussed from the time of foundation of the port of New Orleans in 1717, including its economic development and that of its surroundings in the lower river valley, which has led to an unequal battle between humans and floods. The delta has developed, but nature has retained rights that the current environmental crisis constantly reinforces, so significant has been the reduction in contribution of the river to sedimentary equilibrium of the delta. Today, the question arises of a choice between an endless pursuit of technical proficiency and partial disinvestment, which certain stakeholders propose in view of a return to the *wilderness*.

The last chapter widens its outlook to draw up *an overview and put forward proposals*. The overview is composed of the dynamics and condition of delta structures in the face of erosion by the sea and the rise in water levels at the very heart of deltas. These challenges mobilize science, engineering and defense of the natural state in a context of the growing role played by hydro-diplomacy. This is because the questions of durability or quite simply of delta survival, are those to be asked in the long term. Only vigorous measures examined at the scale of watersheds can guarantee the future of these spaces and of the populations that rely on them, and contribute to the food equilibrium of numerous developing countries.

Acknowledgments

The author extends warm thanks to Yves Bégin, David Blanchon, Geneviève Bravard, Marc Goichot, Richard Marston, Michel Meybeck and Ferréol Salomon for their shrewd advice during the proofreading stage; Thierry Sanjuan for his initial suggestions; Colette Bedoin for her presence during a difficult technical stage.

The author also wishes to express his gratitude to Diane Fremiot (Editions Colin) and Steve Gorelick for their kind authorization for reproduction of material.

1

Deltas: Young, Fragile and Threatened Environments

Deltas are structures formed from sediments of mainly fluvial origin, built by a set of processes that combine the actions of rivers and the sea; they exhibit a set of fluvial and coastal forms and, in general, a morphology that protrudes from a coastline that may be more or less exposed.

Effectively, delta formations are constructed when sedimentary materials of continental origin hold more sway than the destructive action of the receiving aquatic environment; the latter being due, either independently or jointly, to long-shore drift, swell, waves and tides.

Delta formations have developed over the last 6,000 years[1], approximately, in other words since the relative stabilization of the Earth's sea levels, at the mouth of watercourses entering an ocean or a sea.

Fluvial sediments are deposited when water flow coming from the continent, having lost its power when entering the marine environment, is forced to deposit its load. Deltas are developed fully when they extend large fluvial bodies that drain vast watersheds subject to high levels of erosion and that carry abundant sedimentary flows. Conversely, a river with little input of sediment ends with an estuary that is more or less filled in and subject to tidal effects.

1 In principle, they are younger in their distal sections (downstream), which go a long way to explain their fragility in the presence of oceanic currents.

1.1. Long-term construction of deltas: general mechanisms

1.1.1. *Processes and basic forms*

Delta formations generally fall into one of three types of basic formations and sedimentary facies. Regarding the sea level, deltas are organized from upstream to downstream in the following simplified way.

1.1.1.1. *Delta plains or subaerial deltas*

The upper delta plain (channels not affected by salt water) and the lower delta plain (subject to tidal effects) can be distinguished by the limit of influence of the tidal regime; this limit depends on the slope of the delta and the flow rate in the watercourse. The distribution channels* are produced by avulsion*, which creates a new channel when the main channel rises (or vertically aggrades) by sediment deposition, and a flood opens a breach in one of the levees bordering the channel. The levees, made up of sandy materials, are separated by shallow basins that form freshwater or briny wet zones which are rich in ecology. The general facies of the deposits features beds inclined at low angles at the surface; aggradation of the delta plain is slow because materials are transported at the surface towards the delta front.

1.1.1.2. *The delta front or proximal part*

The front combines a narrow subaqueous platform and the "front" itself, which progresses (or "progrades") more or less rapidly depending on the intensity of fluvial input, generally sandy and/or silty. The front has a topographic slope of 10° to 25°, which conforms to the dip of deposits brought down to the mouth.

1.1.1.3. *The prodelta or distal part*

The prodelta is the subaqueous part of a delta that rests on the continental shelf*. The prodelta, created in a zone of deep water with an ocean floor of gentle slope, is made up of very fine deposits, silts and clays from suspended loads carried by plumes. The deposits are made up of laminated beds. This unit is itself fossilized by the progressing delta front.

1.1.2. *Dynamics of construction and redistribution in progress*

The plan form of a delta is conditioned by the interrelation of three competing forces specific to the receiving environment, one fluvial and the two others marine in nature; each of these forces can win over the others depending on the intensity level of its action. This set of influences is the basis for ternary classification of deltas, on a genetic basis, that remains the most widely used [BHA 06, GAL 75].

1.1.2.1. *River-dominated deltas*

River-dominated deltas are subject to the dominant combined action of the liquid and solid flow rates of the river (made up of the bed load* and the suspended material*). These deltas present characteristics of formations undergoing active construction, in contrast to the two following types. They have an elongated shape, whether it is simple or composite, constructed as an extended prolongation of the lower delta plain. The distribution channels or branches end in sandy mouth bars*. The branches separate bays that are progressively transformed into marshes. The Mississippi Delta is the archetype of a river-dominated delta, at least in its northern part, which is constructed along the main fluvial channel (Figure 1.1).

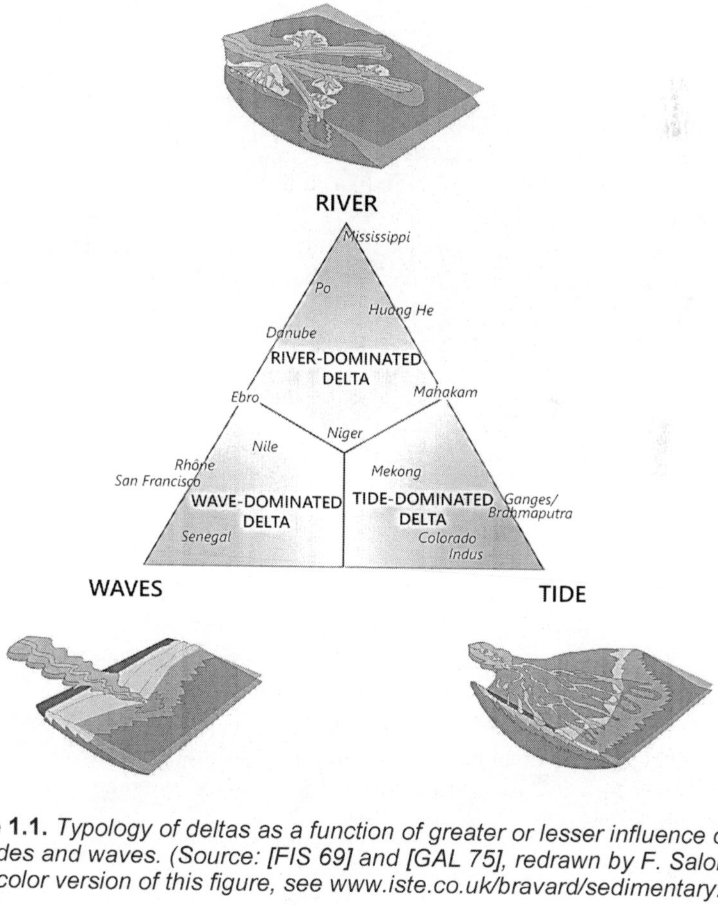

Figure 1.1. *Typology of deltas as a function of greater or lesser influence of the river, tides and waves. (Source: [FIS 69] and [GAL 75], redrawn by F. Salomon). For a color version of this figure, see www.iste.co.uk/bravard/sedimentary2.zip*

1.1.2.2. *Tide-dominated deltas*

These deltas are formed from the branches of distribution channels shaped in a lower delta plain by the action of tides; the branches are called "estuaries*"; the average tidal range is felt upstream and its influence is dominant in comparison with the influence of the river. Delta branches under tidal influence (flood* and ebb*) are the most stable. Delta construction extends slightly or not at all into the sea, apart from with subaqueous mouth bars lying in an extension of the distribution channels and shaped by tidal effects. Between the mouths, vast areas of mud can develop on the tidal flats*, often intersected by thin sandy bars called "cheniers*", and colonized by mangroves which are forested formations adapted to salt water. This type of delta is well illustrated by the Amazon, Ganges and Brahmaputra Deltas, or even by the Yangtze Delta.

1.1.2.3. *Wave-dominated deltas*

The coarsest fraction of sediments carried down by the river (in general, sands that belong in part to the bed load and in part to the suspended load) is redistributed more or less actively by the waves (mainly as a function of their average height) and by long-shore drift*, in sandy bars parallel to the coastline of the delta; these bars or barrier beaches have seasonal openings and may isolate lagoons. Under the action of waves, the deltaic protuberance takes on a lobed form, often pointed in shape. When the deltaic lobe* loses input from fluvial materials, it erodes, regresses and melts into the coastal bars, shaping the shoreline between the active mouths. This type of delta is seen in the Senegal and Nile Deltas, as well as in the Tiber and the Rhône in the Mediterranean.

While this classification is convenient, in practice deltas often take on mixed and evolving forms, in particular by the association of forms reworked by waves and tides, whether the delta is progressing quickly or slowly under the influence of materials of continental origin. This is why certain specialists look at the balance between the suspended load of rivers at their mouth on the one hand, and the relative importance of the action of waves and tides by means of their ratio on the other; this is a means of providing a qualitative basis for grouping deltas together in large families [HOR 07a]. The deltas of the Orinoco, the Red River and the Mekong belong to a mixed type. The Mekong Delta has sandy coastal bars shaped by waves, as well as estuary-type branches in its northern part and mangroves to the south, where muds are deposited downstream of the direction of long-shore drift.

Furthermore, the proportion of sand and mud coming from the continent conditions the formation and future of deltas. Deltas with a high proportion of sand (more than 10% of the total load) have a greater area above sea level than mud deltas; however, the presence of mud is essential because it consolidates soils, makes them more resistant to fluvial and marine erosion and encourages good health in vegetation formations such as marshes and mangroves. The latter two increase the resilience of the delta to the forces applied to it by the ocean [GIO 14].

1.1.3. *Young and unstable areas*

1.1.3.1. *Natural sinking and delta equilibrium*

Natural or geological subsidence is a slow and regular movement that reduces the level of the ground surface. Several processes can exist, sometimes simultaneously:

– on the one hand, a deep process can sometimes be at play, generally related to the movement of the lithosphere, for example linked to downward movement due to fault tectonics or to large-amplitude warping that affects sub-delta sedimentary deposits;

– on the other hand, at a different timescale (of climate periods within the Quaternary epoch) and in the regions of the Earth that have undergone glaciation, these vertical movements can also be isostatic* in nature, and will become more obvious as they approach the poles. Under the weight of ice, portions of continent have sunk by several meters, causing an equilibrium process with uplift of their external marginal areas; this is called an "isostatic adjustment". During melting of ice sheets, the process is inversed: deglaciated areas are uplifted and their marginal zones sink; this is called "isostatic rebound";

– finally, subsidence can be due to compaction of sediments that have been deposited in delta formations. The recent age of deposits (see the later section) means that they have not yet developed cohesion; this develops over time and depends on the nature of the sediments deposited.

The scale of these types of geological sinking, generically described by the term "subsidence", is modest, in the order of a few mm/year. It goes without saying that deltas have formed because the input of sedimentary materials from the continent has exceeded the loss of volume due to subsidence. While the speed of the processes acting in favor of subsidence is relatively slow and always active, and can be predicted and modeled, the (over)compensation imposed by continental materials (which leads to a net positive balance or construction of the delta structure) is faster, fluctuating in space and time. With today's knowledge of the history of the Earth's climates and erosion, we are able to understand the history of sedimentary budgets* in deltas and, to a certain extent, to model their future based on initial hypotheses that combine all the relevant factors in each case.

1.1.3.2. *Rise in sea levels (> 6,000 years BP), initial phase of delta construction on a global scale*

Fluvial valleys were deeply incised during the cold periods of the Quaternary in which "eustatic*" marine regression occurred; the last one had an amplitude[2] of 110–120 m. When the climate warmed, the sea level rose rapidly. Deep deposits were extracted from deposited materials by deep core drilling, which were then analyzed and dated [SMI 11].

A very clear change in the nature and rhythm of deposits took place around 8,500–9,500 cal BP* when the sea level rose abruptly; a very credible hypothesis is that part of this rise is believed, at least around 8,500 cal BP, to be due to the sudden unloading of North American lakes (Agassiz and Ojibway lakes) which had until then been blocked by the Laurentides ice sheet; the enormous volumes stored (163,000 km^3*) led to a presumed effect of several decimeters on global ocean levels over the course of a year [TEL 02]. Deposits associated with the rise in sea levels were estuarian in nature and strongly influenced by the tide, with marine waters penetrating the lower valleys that had been temporarily transformed into rias*.

The Earth's deltas were created when the rise in the levels of oceans and seas that communicated with oceans slowed down; the balance of forces between fluvial material input and the rise in sea levels then reached an equilibrium before tending more towards construction behavior due to input of materials of fluvial origin [STA 94]. The "jump" in sea levels associated with unloading of continental waters, which is itself followed by a phase of stabilization, may even be at the origins of delta formation [HOR 07b]. Stabilization of the sea level, progressively achieved between 8,500 and 6,000 years BP*, is thought to have then allowed deltas to be constructed by uplift and by progradation or extension into the sea. A delta in its first stage of formation at the back of a bay is initially influenced by the tide; once formed, it extends into the ocean and becomes more sensitive to wave action, which is exhibited by the construction of coastal bars. The Po and Rhône deltas have not been subject to tidal influence, but only to the influence of waves and long-shore drift.

If the delta receives a good supply of materials from the continent, the delta plain uplifts and the delta front progrades. Deltas with high levels of fluvial activity (floods, deposition in the channel), such as the Mississippi, exhibit processes of

2 Eustatic glacial variation is in this case the reduction of sea levels related to the retention of water on continents in the form of ice.

channel migration by avulsion as well as deltaic lobes. In contrast, tide-dominated deltas with deep and stable channels undergo avulsion processes to a lesser extent, even though they can possess channels formed during the periods of high activity which were subsequently abandoned.

The deficit of a fluvial sedimentary budget – which has its origins in disruptions at the scale of a catchment* – is likely to modify the deltaic sedimentary budget. The deficit of input materials at the river exit disturbs, in particular, the accumulation on both the delta plain and the delta front. This accelerates the subsidence of the plain, even though the sea level rises under the effect of climate change that is affecting the Earth as a whole.

In addition, progression of the delta front can be slowed down or sometimes cease, and the front may regress. However, the observer is faced with a question of scale because the evolution of a sector should not be confused with the evolution of the sedimentary budget. Sectors of the coast can regress in a delta that is advancing, simply because the sediment cells* have a net negative balance under the action of long-shore drift; other cells benefit from material input from nearby mouths, where this fresh material is carried by long-shore drift. The balance at the scale of the delta cumulates the values obtained in the individual cells, which can thus be positive or negative overall.

1.1.3.3. *Dynamics of behavior of deltas in the Holocene period*

The behavior of deltas over the last 6,000 years differs from one to another depending on whether tectonics is causing a positive or negative movement of the continent at the regional scale, and whether the delta sediment itself is more or less compacted after deposition. While the sea level is rising, the delta regresses under the influence of the tide, and the action of the latter causes the funnel and estuary-type branches to migrate towards the interior of the delta.

The behavior of deltas is also different for different regions of the world as a function of the history of the rivers that have created them. Certain rivers have had small variations in their liquid flow rate and load over the course of the last few millennia; others have had strong reactions to land clearance associated with pastoral and agricultural activities, in which case the Anthropocene* epoch pairs up the mutation of rivers and that of their subaerial deltas. There was a much faster rate of progradation* of the Huang-He and the Yangtze after 2,000 years BP, due to the increase in sediment production* in their basin and perhaps to a reduction in deposition along the Yangtze.

1.2. Some of the Earth's last great natural deltas: two deltas in the Arctic

We will take two examples of Arctic deltas, the Lena in Russia and the Mackenzie in Canada. These deltas have remained very natural, with very little development of their fluvial basin and very low levels of human occupation, apart from a little navigation activity on the main Lena estuary. Very specific processes are at work under the effects of cold temperatures, but we have selected them because their characters have remained very natural.

1.2.1. *The Lena Delta*

The Lena River, 4,400 km long, drains a basin covering an area of 2,490,000 km^2, and has a flow rate of 16,300 m^3/s at its mouth. Its hydrological regime is characterized by a maximum in spring (60,000 m^3/s) and a low water period in January (3,000 m^3/s).

The delta, constructed in the Laptev Sea, has an area of 29,600 km^2 and is the largest in the Arctic regions. The permafrost extends below an active layer, 30–50 m thick, which alternates between freezing and melting each year. To the west of the delta, several pre-Holocene thermokarst* terraces have been lifted up by tectonics and slowly eroded by thermoabrasion* and thermoablation* (with the oldest terrace being at an altitude of +20 m) (Figure 1.2).

The active deltaic landscape in the Holocene is formed from several very distinct lobes. Recent work has shown that in contrast to publications about the Earth's deltas, Holocene deposits of the Lena Delta have been laid down in the last 8,000 years, especially under the influence of the sea, and more precisely during phases of marine transgression* (thick organic levels); fluvial deposits, forming thin layers of silts and sand, appear less here.

Organic deposits, quite different from peat, were laid down during summer in pools of standing water blocked during the phases of rising sea levels. Organic deposits are composed of mosses and sedges from the sweeping action of the repeated passing of marine waters across the coastal tundra.

The process of blocking estuarian waters by rising sea levels has occurred in several multisecular periods of time, for example around 2,500–1,500 years BP and 400–200 years BP; during the periods of marine regression, the deposits were partially eroded and the material was carried into estuaries that became active again [BOL 15].

Figure 1.2. *The Lena Delta (Siberia). Infrared channel image taken on 27 July 2000, during unfreezing, by the satellite Landsat 7. We can distinguish the active channels that feed the eastern part of the delta (purple), the recently abandoned channels in the center (light green), the deltaic plain (dark green) and the Holocene levels raised to the west with lakes nestling in the thermokarst depressions and oriented in the direction of the prevailing wind. (Source: NASA Observatory and USGS ROS Data Center). For a color version of this figure, see www.iste.co.uk/bravard/sedimentary2.zip*

1.2.2. *The Mackenzie Delta*

The Mackenzie (1,738 km long) drains a watershed of an area of 1,810,000 km^2, and has a magnitude of 9,700 m^3/s. For this river in the Northwest Territories of Canada, high water occurs in June when the snow melts (21,500 m^3/s on average), and a low-water season lasts through the winter from December to April (3,400 m^3/s on average). The delta covers an area of 13,500 km^2 and has sediments that are 70–80 m thick; they rest on the substratum formed from Precambrian and Devonian rocks overlaid by Cretaceous layers with high potential for oil and gas exploitation. Below the surface of the delta, except for lakes and channels, the permafrost* holds its place. Global warming at the end of the 8th millennium BP allowed for the formation of aquatic plants and peat bogs, before a period of cooling that favored the development of permafrost, and later slowed down the formation of peat [VAR 97].

Today, the water surfaces are frozen for 8 months of the year and covered in snow. Despite the harsh climate, but thanks to the proximity of the northern edge to the boreal forest and the presence of water, biological diversity is high, which is the subject of a conservation policy. The geography of 25,000 lakes, covering half the surface of the delta, constantly changes as the active channels are eroded, fluvial sediments are brought in, the permafrost melts or fluvial branches are abandoned; it also depends on the hydrological balance between inputs and outputs (Figure 1.3).

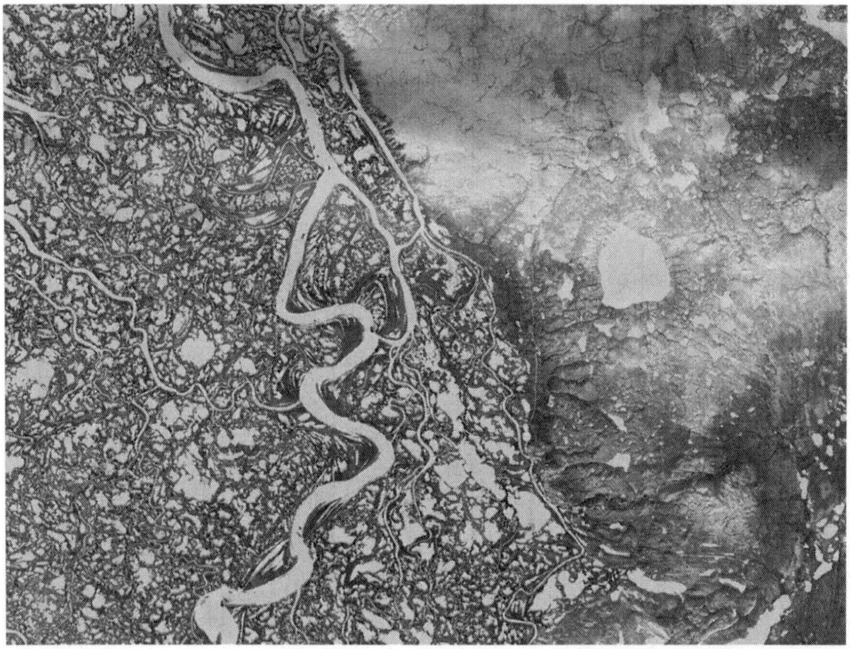

Figure 1.3. *Details of the Mackenzie Delta (Northwest Territories, Canada). This view, dated 16 November 2016, shows the eastern branch (East Channel) and the thermokarst lakes overtaken by ice. In the warm season, the river overflows but deposits few sediments. (Source: © NASA Earth Observatory, image by Joshua Stevens, using Landsat data from the US Geological Survey, available at the link: http://earthexplorer.usgs.gov/). For a color version of this figure, see www.iste.co.uk/bravard/sedimentary2.zip*

Lakes are defining features that characterize Arctic deltas, but their low levels of organic production do not cause them to be filled in. The majority of exchanges between the channels and the lakes (lake flooding) occur during the period of high waters in the month of May, whereas the lake levels are lower in summer because evaporation exceeds the input from rainfall. The frequency, intensity and duration of exchanges depend on the relative altitude of each of the lakes, an altitude that itself results from multiple influences (fluvial flow rates, logjams and debacles, variation of sea levels). Hydrologists believe that the alteration of fluvial flows by the construction of dam reservoirs would affect the levels, lake flooding and the high productivity level of the zone [MAR 88, MAR 89].

1.3. The Earth's deltas: what is their current situation in the face of terrestrial and marine constraints?

The rise in sea levels is a reality that combines several factors; it is modified in each of the Earth's deltas and even within each delta. Based on the results obtained for 40 deltas around the world, in terms of the predominant factor, the main causes are:

– the reduction of input materials of fluvial origin, in itself due to sedimentary retention and the consumption of water on the continent: < 70%;

– the extraction of resources, a factor that accelerates subsidence: 20%;

– the rise in sea levels: > 10%.

At the current rate of sea level rise, from now until 2050, specialists believe that 8.7 million inhabitants will be affected by flooding across about 5% of the total deltaic area studied [ERI 06].

1.3.1. *The rise in sea levels*

The rise in sea and ocean levels, estimated for a long time using tidal marker posts, was found to be 1.2–1.9 mm/year during 1900–1990. A precise estimate has been made possible at all spatial scales since the use of high resolution imagery with good stability, from the TOPEX-Poseidon satellite (1992), and subsequently from the Jason satellites, with a temporal resolution of 10 days. The average rise during 1992–2015 increased to 3.3 mm/year on average (incorporating variations over a short time step) [FAS 16].

The "rapid" rise in sea levels has become a social issue of primary importance since the release of the 4th Intergovernmental Panel for Climate Change (IPCC) report in 2007, and the certainty that the rate, which was 1.8 mm/year in the second half of the 20th Century, has now reached 3.1 mm/year or more [SOL 07]. The rise is considered to be rapid when it exceeds a rate of 3 mm/year; the reasons for this are the warming of the oceans (+0.64 $W.m^{-2}$ between 1993 and 2008), the volumes of which increase due to dilation, and the quantity of water input, due to the reduction of continental ice mass by melting and sliding into the ocean (more than 500 Gt/year*) [CHU 11, HAN 10].

In general, the global rise in sea levels is taken into account, but here we will examine the regional conditions, insofar as the very rapid melting of continental ice causes differential readjustments of the Earth's surface. These include "glacio-isostatic adjustments*" (at the scale of millennia) and "elastic deformations" of the Earth's crust that is lifted rapidly and immediately when it is freed from ice sheets, such as in Greenland and in the Antarctic (the rise in sea levels being reduced by a corresponding amount, but reversibly so); the rise due to elastic deformation can be in the order of 1 mm/year. The regional differences or anomalies also include the differences in water temperature (the western Pacific has warmer waters than the north-west Atlantic, for example) as well as the circulation of masses of oceanic water of greater or lesser degrees of salinity (spatial variation of –4 to +4 mm/year). Warming of the oceans at depths between –50 and –300 m also plays a role in the melting of glacial fronts in contact with waters, which is still not fully understood. All things considered, the rise is more rapid in the oceans of the Southern Hemisphere than in the high latitudes of the Northern Hemisphere (Arctic coasts and Alaska where the continent is being lifted). A summary of recent publications leads to an estimate of a rise of at least 10 mm/year in the decades to come [CRO 12].

However, storage of water in artificial reservoirs at the surface of continents (10,800 km^3 in total) is believed to have reduced the rise in sea levels by 0.55 mm/year over the last 50 years (therefore a total of about 30 mm), or 2.46 mm/year over the last 80 years [CHA 08]. However, this is without counting the effect of exhaustion of underground reservoirs to the benefit of surface flows and the atmospheric balance of the water; the negative balance of the use of underground water caused sea levels to rise by +0.57 mm/year around the year 2000, and this could exceed +0.80 mm/year around 2050. The net balance of the effect of reservoir storage and the effect of released water extracted from underground reservoirs was therefore –0.15 mm between 1970–1990 (a period of increased reservoir construction). It was positive at +0.25 mm/year between 1990–2010 due to the reduced rate of dam construction and to the accelerated rate of extraction; the rate could rise to +0.84 mm/year by around 2050 [WAD 12].

The rise in sea levels can be accompanied by risky situations. The coastal communities of the United States (East Coast and Gulf Coast) could justifiably live in fear of worsening storms and flooding related to exceptional tides:

> "Around the year 2045, so the time it takes for the mortgage on a house to be paid back, more than half of the 52 communities analyzed could see the risk of flooding by the tide increase 10-fold or more, and a third of these communities could be subject to more than 180 flood events of this type each year" [SPA 14].

Due to being lifted by 30 cm over a period of 30 years, communication channels outages are a current feature of life on many American coasts.

1.3.2. *Sedimentary exhaustion of continents*

The global sedimentary budget has been updated on the basis of numerous international publications (Volume 1, Chapter 4) (Table 1.1).

Sediment balances at river outlets	Billions of tons per year
Reconstituted geological or prehuman flow	15.1
Part of the flow stored by reservoirs	3.4–5
Reconstituted current flow, excluding the impact of reservoirs	16.2–17.8
Current flow	12.8

Table 1.1. *Sediment fluxes on the Earth's surface*

These figures show the scale of the reduction of fluvial input to the oceans in the last few decades. This evolution weakens the deltas in the face of the effects of dynamic processes at play, by creating an imbalance of their sedimentary budget at the scale of the deltaic plain and front. An evaluation on a global scale reveals a total trapped volume of 73 km^3, a volume that will not reach the oceans [SYV 11]; by supposing a deltaic progression wedge of an average thickness of 10 m, this corresponds to an area lost due to the effect of trapped material in artificial reservoirs of 7,300 km^2 for deltas around the world.

Asia provides a very large contribution to this reduction because its nations have chosen to follow an accelerated path to development. They have chosen to convert part of their carbonated energy production by focusing on green energy to design their new energy mix, while not giving environmental concerns the place they deserve.

A large summary [GUP 12] was carried out on the impacts of mega-dams* in East and South Asia (Table 1.2).

Watercourse	**Pre-dam flows (Gt or billions of tons)**	**Current flows modified by dams (Gt)**
Ganges–Brahmaputra	1.67	0.850
Indus	0.110	0.037
China *Huang-He*	1.8 *1.2*	0.415 *0.15*
Indian peninsula		0.083
Southeast Asia	0.572	0.81
Total		2.15

Table 1.2. *Reduction in sedimentary flux rates in East and South Asia by mega-dams [GUP 12]. The flow rate previously estimated was 4.74 ± 0.7 Gt [SYV 05]*

The sedimentary budgets can be specified at the scale of basins, as done in South Asia (Table 1.3). It is estimated that in this region, around the year 2050, nearly 9 million people and 20,000 km^2 belonging to 33 different deltas will suffer from flooding and coastal erosion aggravated as a result of input materials of continental origin which are incapable of compensating for subsidence and the rise in ocean levels.

Delta	**Percentage of reduction of the fluvial load**
Indus	94
Ganges–Brahmaputra	30
Krishna	94
Narmada	95
Cauvery	80
Sabarmati	96
Mahanadi	74
Godavari	74
Brahmani	50

Table 1.3. *Percentage of reduction of the load carried by the rivers of South Asia to their deltas over the course of a century [DAN 14a]*

1.3.3. *Extraction of resources and accelerated subsidence of deltas*

Globally, 300–500 million inhabitants live in deltas threatened by the effects of subsidence or simply susceptible to be affected by them at some point. And 20% of the Earth's deltas have accelerated subsidence, which is a primary cause of the relative rise in sea levels; accelerated subsidence must be understood as an aggravation of the process of natural subsidence by human activities. In the event of flooding, the very low altitude of the ground, at times negative, necessitates temporary displacement of the population, such as in Jakarta (200,000 people in 2007), with epidemics associated with mixing of contaminated waters with underground waters or flood waters. Pumping facilitates intrusion of salt water into underground waters, and the underground networks that transport pollutants can rupture [SCH 15].

Earlier, we saw that geological subsidence is a slow and regular lowering of the ground surface, observed in the regions of the Earth where relatively recent sediments undergo compaction processes; they can also be carried into a depression in the lithosphere that they cover. This natural movement is minor when compared to certain dynamic processes in progress. Under what mechanisms are they operating? Deltas can be areas of the Earth that contain natural riches, such as freshwater and hydrocarbons, trapped in layers deep underground. Exploitation of liquids results in a worsening of natural subsidence, but this time at rates that have no comparison to natural rates, in particular in Southeast Asia, nor with that of the rise in sea levels.

This is due to the following reason: pumping activities extract water from the pores present between the unconsolidated particles, and thus the pressure decreases, which induces compression and subsidence of the surface. An estimate made for the Bangkok plain produces a loss of volume at the surface of 0.05–0.1 m^3 for each cubic meter pumped at depth; in other words, daily pumping of 2 hm^3* should be compensated for by daily backfilling of at least 100,000 m^3 of earth; without this, the negative effects of subsidence are accentuated [PHI 06]. Subsidence would not be too serious if the pores between the grains could go back to their original volume when the water returns, in which case the ground would rise; the water table itself may rise if extraction of water or other fluids are abandoned, but its return to equilibrium would place it above the level of the soil because the latter does not rise.

1.4. Subsiding deltas in Southeast Asia

1.4.1. *An example of a young, mainly rural delta, the Huang-He*

One of the world records for accelerated subsidence, if not the record itself, has been found in the Huang-He Delta, where, in order to provide water for a young aquaculture establishment, pumping has caused local lowering of the ground surface by 25 cm/year [HIG 13]. The values reached fall into those monitored in some delta cities, therefore with an extreme range for urban areas. This delta is itself very young, since it has been in formation in the Bohai Sea since 1855 on the most recent mouth of the Huang-He, which we know to have been subject to avulsions over the course of its history. It has a surface area of 5,500 km^2 and a thickness of 15–20 m if just the contemporary sediments are taken into account. Chapter 4 of the first volume showed that the construction of dams caused the flow of water reaching the single lobe of the Huang-He to decrease from 43 to 4.9 km^3/year [HIG 13] and the sedimentary flux from over 1 Gt to 0.15 Gt/year.

Pumping of briny water to reduce salt levels in the water of shrimp farm containers, installed between 1970 and 2000, has reached the level of 1 km^3/year; at the same time, deep oil exploitation has developed, although its effects have not been clearly demonstrated up to the present day. The impact of these practices is considerable, as radar images have shown. Despite the construction of sea walls, the north coast has receded by 7 km, including the delta mouth area, even more quickly since the erosion is attacking a lower coastline. The shrimp hatchery area, affected by the peak in subsidence, has a diameter of more than 2 km. In 4 years of measurements, subsidence exceeded 5 cm/year in an area of 70 km^2. The high levels of subsidence are explained by the very high porosity of the clay and the uncompacted silt (40–55% voids).

1.4.2. *Urbanized deltas in Southeast Asia*

We will observe that the most serious situations are found today in deltas that have undergone significant urban development, since some of the world's megalopolises are located there. These are known as *sinking cities*. The distinction between how much is due to the rise in sea levels and how much is attributed to the acceleration in sinking due to the extraction of resources and the weight of the built environment is not clear; more precisely, converging effects are generally combined within the same dynamic. It goes without saying that New York, on the Hudson estuary, is the epitome of a sinking city in the eyes of Americans, but here the rise in sea levels takes precedence over subsidence[3]. We have chosen large cities in Southeast Asia.

3 "If there's any place in America that we're probably going to try and defend (from the sea), it's Manhattan and New York" [JOH 13].

1.4.2.1. *Kanto, Japan*

The Kanto region, home to the capital of Japan, Tokyo, has been particularly affected by water pumping [FUR 15]. The groundwater resource is found in quaternary sediments over a thickness of 2,500–3,000 m in the coastal basin. In addition to this extraction near the surface, natural gas is also extracted from the deep, fossilized marine beds. Subsidence began in the 1900s below Tokyo and spread around Tokyo Bay in the Koto area; it had already reached a cumulative level of 1 m in the 1930s but ceased when activities stopped during the war; it took off again sharply in the years 1950–1955 when industrial demand was 1.25 hm^3/day for the 40,000 factories in the region (industry uses 80% of extractions). Peak subsidence was even 24 cm/year at Edogawa-ku at the end of the 1960s due to the growing depth of wells bored in the plain (200 m in the Edogawa and Koto districts). The affected area was 290 km^2, and the cumulative sinking reached 4.5 m in industrial districts. Recent subsidence tends to migrate towards the north of Kanto since the extraction peaks have also moved.

Awareness of the issue came to the forefront at the beginning of the 1960s, with the risk of marine submersion by typhoons becoming an even more worrying reality since the rupture of dykes is likely to be accompanied by the submersion of areas that are now below the average sea level. The policy of extraction control only came into effect from the 1970s. It has been effective, sometimes to the point where the rise in water levels has locally affected cellars that were built during the phase of maximum reduction (sometimes + 30–40 m of rise for a maximum reduction of the water table of 60 m). It should be noted that demand is considerable for the 40 million inhabitants now living in this region, for its industrial waterfront and for agriculture, but it is acknowledged that human needs are now considered to be a priority; these are potable water and supply to air conditioning devices. Industry has had to resort to rivers and treatment of used water. The pumping of potable water for the population of Tokyo has been reduced to 0.5 hm^3/day, with very careful monitoring of water quality, which has in fact decreased. Underground reservoirs are also resupplied with surface rainwater flows. A clear result; in all places, the current rate of subsidence has returned to values below 2 cm/year [SAT 06].

1.4.2.2. *Bangkok, Thailand*

The capital of Thailand has 12 million inhabitants; located near the mouth of the Chao Phraya (a river with an average annual flow rate of 880 m^3/s), it covers more than 550 km^2 of a basin that has an area of 160,000 km^2, and that itself covers nearly one third of the country. Under the basin, the layers of the Tertiary and the Quaternary are 500 m thick, while some are aquifers that are separated by layers of clay. A major concern is how to distribute the water resources in a basin that suffers from a shortage during the dry season. With a low population in the middle of the 19th Century, the delta became a rice-growing area after installation of the capital city at Bangkok in 1767 and, above all, from the 1860s onwards. Competition for water is intense between agriculture in the Chao Phraya basin and exploding levels of urban demand. Thailand has taken advantage of the monsoon to store water in very large reservoirs in foothills, namely Bhumibol (1964; 13.5 km^3) and Sirikit (1974; 9.5 km^3), and by using it for rice irrigation. Based on this geographical distribution, it is logical that Bangkok has turned to underground water, abundant and of better quality than overused surface waters (the pumping rate being 36 m^3/s in 2006). However, in the context of local tensions, it is essential for the fluvial estuary to be crossed by a high flow rate of freshwater to limit salinization of the water by the upstream movement of the salt wedge, even though shrimp cultivation in the lower delta is not at all compatible with the high levels of pesticides used in rice cultivation [MOL 06].

Subsidence of the Bangkok plain came to light at the end of the 1960s. It has two major causes. The first is an excess of water extraction: pumping, mainly private initiatives for industry, occurring at a rate of 1.2 hm^3/day at the beginning of the 1980s and 2.8 hm^3 in 1998, before reducing to 2 hm^3/day. The second cause of overloading is the compacting of a thick layer of soft clay present on the surface by buildings and road infrastructure. Bangkok has, for example, 700 buildings with more than 20 floors and 4,000 buildings with 8–20 floors. Since the level of the overexploited water table has decreased by 70 m, subsidence is active. It began in the 1970s and reached a peak at the beginning of the 1980s, with a rate, fortunately brief, that exceeded 100 mm/year. However, its cumulative value has remained below that of some deltas, with a maximum of 2 m. General movement has clearly diminished since then thanks to measures to restrict consumption, but the surface of the affected zone is spreading. It affects, for example, the area around the Bangkok airport, with a cumulative value of 70–75 cm [BHA 13]. It goes without saying that sinking of the soil surface is a great inconvenience, because it causes damage to buildings and streets and because it is a great handicap for the management of fluvial floods. The layer of clay, which gets thicker in the deltaic part of the plain, effectively retains floodwaters in the basin depressed by subsidence.

The original site of Bangkok, created in 1782, is located on a levee of the Chao Phraya; since then, the city has extended down into the swamps, where the river flows next to a distributary* (the Tha Chin), a tributary (the Pa Sak) and many drainage channels. However, the fluvial network is not suitable for large floods, even though flooding is limited, in principle, by storage in the upper basin. The hydrographic network has almost no slope, and the rise in sea levels presents a problem for the evacuation of terrestrial waters. The roads, raised little by little as the marshes sink, railways on embankments, and the small dykes in the fields divide up the plain and constitute obstacles to flows when the basin is exposed to flooding, and the waters take weeks to dry off. Central districts are protected by high dykes, such as the Royal Dyke (constructed after the 1983 floods), drainage tunnels and locks that open out into the Gulf of Siam; some districts have even been polderized since water pumping has been permanently there.

The serious 2011 floods took place during a long rainy period, which lasted from May to October and produced cumulative precipitation of 1,700 mm over the capital city. Overflows began upstream from the plain of the Chao Phraya in July, and the floods reached the northern suburb of Bangkok in October, after the rupture of dykes in several places. Resistance of the town authorities to the principle of allowing water to pass – the town is a narrowing in the evacuation of the river towards the ocean – has created quite violent conflicts with the flooded suburbs. The town authorities, whose dykes of nearly 80 km in length were a dam for the flows coming from the basin upstream, were ordered by the government to allow the waters to pass through to the sea, even though this meant inundating the city. The rise in phreatic waters and the cumulation of rainwaters contributed to inundation of areas that were in principle protected. The 17 km^3 of floodwaters that were not retained by reservoirs did not totally recede until February 2012, with sea locks and pumps only able to evacuate 1 hm^3/day.

How could the sedimentary input coming from the basin balance out subsidence? A series of close-up satellite images taken in 2011 and 2012 show the path of floodwaters, as well as the relative water heights; they give information about their turbidity and show that sediments reach the plain through breaches created in the levees. They tend to be deposited close to former channels, but the network of artificial obstacles makes it impossible to spread them across the plain [LIE 16]. The load of the Chao Phraya strongly correlates with the flow rate, as shown by the values measured during the 1995 and 2011 floods, which are higher than values before the construction of reservoirs. The effects of clearance work that has been converting the natural vegetation into agricultural lands since 1960, and that has increased surface flows, have doubtlessly more than compensated for the impact of the reservoirs. The load transported during the 2011 flood was 28 Mt* (compared to an annual average of less than 6 Mt/year over a period of 60 years), but the technical measures put in place after the 1995 flood (dykes, upstream storage basins) must have reduced its transport towards Bangkok. In the end, the amount of solid material carried into the delta has reduced and cannot compensate at all for sinking of the plain [BID 17].

One of the major effects of the reduction of sedimentary input is seen on the coast. The delta of the Chao Phraya has advanced by an average of 1.5 km^2/year for the last two millennia, but coastal erosion has carried it away for around 40 years, with the regression reaching more than 1 km in 2005. The primary responsibility for regression is attributed to subsidence (1 m in 50 years) due to the reduction of sediment input of continental origin; the latter is also due to extractions of sand in the river estuary (but measurements do not exist). Recession of the coast has exhibited itself with a severe loss of mangrove areas (down from 140 to 20 km^2), since this loss is caused by the development of shrimp and clam farming pools. Deepening of the coast in the intertidal zone* following subsidence increases the energy of waves hitting the coast, and it has been possible to confirm that subsidence of coastal depths by 10 cm is sufficient for muddy coasts to be eroded [SAI 07].

The 2011 high water levels and flooding were disastrous for the population and the economy of the Bangkok metropolis, which were not so much due to the speed of the flooded waters but to their height, and above all, due to the time period during which they stagnated, which is exceptional on Earth. The 2011 flood caused more than 800 deaths around the country, created tens of thousands of displaced persons and significantly affected the economy of the country, with official estimates stating a cost of more than 100 billion euros. The flooding destroyed or damaged millions of dwellings, blocked traffic circulation (bridges became refuge car parks) and required airports to close (including the Don Mueang airport and the Suvarnabhumi International airport); it also paralyzed the production of industrial zones for several months, such as that of Navanakorn. Affected economic sectors were textiles; spare parts destined for the automobile industry of Japanese companies in Malaysia, the United States and Canada; and even the electronic chips used in industry in several countries. Industrial production chains were broken, and the authorities were worried about the confidence of investors in the future of the Bangkok economy.

Construction of a sea dyke is planned for a cost of 15 billion dollars, but certain particularly alarmist experts believe that the city of Bangkok may disappear by 2030 if new reservoirs and a more effective pumping system are not constructed [KOM 12].

1.4.2.3. *Jakarta, Indonesia: the Great Garuda project, the epitome of large hydraulic infrastructure*

The city of Jakarta, Indonesia's capital, with a population of 1 million inhabitants in 1930, today has nearly 10 million concentrated in an area of 660 km^2, to which 18 million inhabitants living in the wider agglomeration must be added. This small port of a Hindu kingdom became a Portuguese trading post in 1522, which was then under the control of the Netherlands under the name Batavia (1596). This port on the island of Java, which was established in 1619, has become the second largest megalopolis in the world after Tokyo. The city presents a striking contrast between the affluent suburbs, constructed in an amphitheater of hills, and the lower city; an area of slums, or at the very least precarious housing. The city authorities have recently razed the Sunda Kelapa district, which was the former Dutch port, after brutally expelling the inhabitants who were pushed back into a far-off suburb; this is just one example among many others of the policy of "urban cleaning" that was previously implemented by the governor Basuki "Ahok" Purnama (now deposed), a policy that can also be seen in the district of Bukit Dori, crossed by the Ci Liwung River channeled between concrete walls. The idea is to support, by means of real estate operations, grandiose projects of *reklamasi* (an Indonesian word derived from the English word *reclamation*) that are in progress in the bay [PHI 17].

As with many cities in Southeast Asia, Jakarta is sinking. Subsidence in Jakarta is the fastest of all large cities constructed on a delta. It causes damage to houses and disruption to the flow of fluvial waters (even causing a change in the direction of the flow) and of used waters (pipe ruptures); it is also responsible for the intrusion of salt water and coastal flooding. Subsidence results from three main causes: the pumping of potable water, the weight of the city and natural compaction of fine sediments carried down by small coastal rivers [ABI 15]. The ground surface has been sinking by 1–4 cm/year on average between 1974–1991; this rate increased from 3 to 10 cm/year in the period 1991–2010, varying as a function of the location and the years, and locally it is 20 cm/year. It is much higher than the rise in sea levels, which is on average 0.32 cm/year (3.2 mm). These two figures are added together. Jakarta holds the world record. The main cause is pumping of underground freshwater at a rate of 180–250 hm^3/day, an enormous quantity that is explained by the needs of large apartment blocks; the latter also have water of excellent quality and escape the cost of polluted surface waters treatment, and the fact that these waters, even treated, are mediocre. Another result of deep overpumping is that poor districts and certain districts inhabited by the middle classes have water that has been degraded by the addition of salt water from the sea, and that they are obliged to draw up from shallow depths, because they do not have access to the highly fragmented network of pipes supplied by private companies; this water is contaminated by floods and used water [FUR 17].

The average cost of flooding in Jakarta has been estimated at 270 million euros/year in an environment where increasing population density increases vulnerability, the exposure to risk (via localization of the stakes) is high and the physical factors evolve unfavorably [BUD 15]. The risk cumulates the effects of flash floods of coastal rivers and the threat posed by the sea. The input of 13 coastal rivers (the largest being the Ci Liwung) that flow into the Bay of Jakarta and the Sea of Java cannot compensate at all for subsidence in a highly urbanized environment that has no capacity to retain waters nor sedimentary deposition processes. The input into the ocean is made of black and foul-smelling clay, and the fluxes of carbon and nitrogen are very high, as are the fluxes of heavy metal. This is because 20% of urban effluents of all kinds exit into the hydrographic network. The erosion of slopes, the impermeability of soils and infilling of marshy depressions have increased the fluxes of water and waste towards the watercourse in such a way that the frequency of flooding has increased. After the tragedy of 1997, the 2002 floods caused 60 deaths and forcibly displaced 360,000 people (450,000 people in 2007 with half the city flooded); in 2013, the dyke constructed by the Netherlands along a flood evacuation channel broke and the waters inundated districts of the city center. If dredging, non-existent between 1970 and 2010, had not been carried out since 2012, the network would have been clogged up. At that time, the World Bank financed the extraction of 4,000,000 m^3 of mud and thus reduced the risk of flooding, but its directors emphasized that the city must introduce what is known as maintenance mentality, which includes stopping rubbish and cumbersome objects from being dumped into the network [COC 15].

The deltaic plain of Jakarta will continue to sink in the years to come, because the water extracted by pumping is considered to be necessary for economic and urban development, even though substitution resources are not available. Only knowledge of the process has made considerable progress. According to specialists, Jakarta, which has already sunk by 5–6 m, could sink by as much again from now to the end of the 21st Century. The rise in sea levels is added to this sinking; in the Bay of Jakarta, it has been 3–4 mm/year on average since the beginning of the 1990s. The threat of worsening intensity of rainfall and extreme flows influenced by the effects of climate change is added to this evolution. An enormous sea dyke is recommended to protect the population that lives below sea level, which today amounts to 4 million people.

Across an area of 2,700 ha, 17 industrial and residential port, polders including parks and water features, have been built or are planned in front of the concave coastline* that opens out into the bay and is protected by a dyke. Based on the power of Dutch hydraulic expertise, they each have an external dyke, a drainage device and pumps to control the level of the water table; pumps have a total capacity of 90 m^3/s to evacuate rains from events with a return period of 25 years and infiltrated waters; they also include partial infilling of polders with sand. This is the case of the Pluit polder, created for residential purposes with its three compartments that cover 220 ha, its body of water intended for water storage and leisure activities. However, both the rise in sea levels and subsidence weigh heavily on this type of construction, considered perhaps a little too quickly to be sustainable. At the scale of the agglomeration, difficulties are presented by the constantly increasing surface flow, the lack of financial means and the lack of coordination between public services for a Drainage Master Plan that has remained incoherent, and has even been considered to have "disintegrated"; at the scale of polders, floods are caused by pumps breaking down, such as in January 2013 [GUN 11]. The recommendations concern delocalization of companies with high levels of consumption to areas outside critical zones of the subsiding plain, as well as infill of topographical depressions when they are not covered by informal housing. The situation becomes very serious. In addition, a new protected city is rising out of the current city in the shape of the Great Garuda[4] project (Chapter 5), but, in the meantime, at the current rate of sinking, the coastal wall will be surpassed by sea levels before 2030 and perhaps before this; in which case, the city center will be flooded more than 6 km from the coast [KIM 17]. This constraint, among others, obliged the company KuiperCompagnons to adapt their project, which won the call for tenders.

1.4.2.4. *Shanghai in the Yangtze Delta, China*

The Yangtze Delta covers an area of 210,000 km^2 and houses a population of 156 million inhabitants; Shanghai accounts for 23 million alone (its population being 1 million inhabitants around 1900). Subsidence began in the 1910s and developed rapidly at the end of the 1970s and at the start of the 1980s, due to an excess of water extraction, then due to an intense pace of construction, which took place here as well as in many other regions of the world. Cumulated subsidence, including natural subsidence, has been recorded since the beginning of the 20th Century in the Shanghai region. It is already 3 m (a little less than the 3.30 m recorded in the region of Tianjin, to the east of Beijing); the area affected by subsidence of 20 cm covers 10,000 km^2 [DAI 16]. Pumping greatly reduced after the peak in the 1950s (did politics play a role?), but control began in earnest in the 2000s. Subsidence then reduced to an average value of 1.3 cm/year and subsequently reduced further, with areas where subsidence is just 1 cm/year today covering no more than 195 km^2. The volume extracted since the 1996 peak has rapidly decreased in comparison with resupply, the volume of which remains more or less constant. The Chinese authorities believe that subsidence in the Shanghai region caused direct and indirect economic losses of 1.7 billion euros between 2001 and 2010, and 38 billion euros since 1921.

Subsidence of the delta is accompanied by significant incision of the southern branches of the Yangtze, with muddy beds and very little cohesion, in the subaqueous part of the delta; the incision has reached the underlying Pleistocene sands. This evolution is due to the significant reduction in sediment input. The load of the Yangtze at its mouth was 240 Mt before 2,000 years BP (considered "prehuman"); today, it is only 130 Mt at the mouth of the delta, having reached higher values in relation to the agricultural enhancement of the basin. The solid flow rate at its point of entry into the China Sea, taking into account the sedimentation in the least active estuarian branches and the marshes, is less than 100 Mt. The reduction was softened by filling the Danjiangkou Reservoir and other reservoirs, by the protection of lands in the upper basin and finally by the Three Gorges Reservoir. The rise in sea levels (3 mm/year) and the erosive action of long-shore drift, which

4 This project was financed by the government of the Netherlands and associated with the Deltares network.

is twice the continental input, also explain a coastal recession of 6 m/year in the unprotected zones [LUO 17].

1.4.2.5. *Hanoi, Vietnam: the Red River*

Hanoi, the capital of Vietnam, has been constructed in the Song Hong Delta (also known as the Red River Delta), that pours into the Gulf of Bac Bo, an appendix to the China Sea. The city's population exceeded 3 million inhabitants and that of the agglomeration is 7 million. Hanoi has problems similar to those of the large Asian cities constructed on a delta.

The Song Hong has its source in Yunnan and drains a basin of 155,000 km^2. The delta is influenced by the waves and the tide, to which the parallel dune bars and the very blunt form of the headland bear witness to this, but the capital has developed upstream from the fluvial part in the western plain. Holocene deltaic deposits have filled in the basin with thick Neogene and Quaternary sediment over a depth of more than 3,000 m. The surface of the delta plain is intersected by sinuous channels, powerful levees formed around 6,000 cal BP and colonized for 2,500 years, and marshes. Reduction of the level of the sea and the river around 4,000 years BP caused a reduction in sedimentation in the deltaic plain, which has allowed the mouth to grow rapidly [FUN 12]. The Red River has a liquid flow rate of 23,000 m^3/s during high waters and 700 m^3/s at low water levels at the end of winter. The solid flow rate of the river, 100–130 Mt/year, circulates during the summer monsoon; it travels between the levees and the dykes without being able to supply the alluvial plain.

Water has been pumped from underground reservoirs since 1909, with a very high demand for it. Since the end of the 1970s, pumping is responsible for lowering of the top of the water table under the city, and for significant subsidence, occurring at the rate of 2–3.5 cm/year; since the beginning of the 1990s, subsidence has caused disorder to communication routes, factories and habitats. During periods of rain, drainage has been made difficult, affecting the streets [NGU 95]. Recent measurements made from satellite data show that the districts on the south bank of the river are lowering at an extreme rate of 6.8 cm/year. The city's surface area is increasing, the construction of large apartment blocks weighs on the alluvial deposits, and it has been shown that subsidence is spreading along with the growth of the city. Moreover, impermeability of the soils reduces recharging of underground reservoirs by infiltration [DAN 14b].

The risk of flooding is partly, but not only, due to the problem of subsidence. High water events and flooding are managed by actors who still lack coordination, despite the approval of a Water Law (1999), whether for management of the river, dykes (monitoring and maintenance), irrigation authorities or major transport routes. The deltaic zone is a juxtaposition of rice paddies, originally linked to the river by gravity (when the river is at a low water level) or pumps (when it is in flood). Today, the paddies have been converted for the benefit of other economic uses and have been partially filled in, which reduces the storage capacity of those that are flooded. The 1945 and 1971 floods have a return period of 100 years and 125 years, respectively, but flood defenses have not been tested at this hazard level since 1971. However, the risk posed to the paddies has been reduced by the increased height of the dykes, to the point of protecting them against a hazard with a return period of 250 years. However, the risk of rupture is very much present, whereas vulnerability is much greater than in the era of rice growing. Selection of paddies to be flooded by spillways, the purpose of which is to relieve the agglomeration, is impeded by the reticence of local authorities to act in the absence of decisions that must be made at the highest level [GIL 06].

1.5. Conclusion

The Earth's deltas are extremely varied and cannot be compared to one to another, due as much to their morphology as to their occupancy levels by humans. The oceanic environment is a feature that unites them, at least by the base level that it sets, if not by the intensity of factors acting to rework alluvial materials through waves and swell. Each one is under the control of a complex, changing, continental influence, with time steps and an intensity that are particular to each of them. However, the common characteristic of the Earth's deltas is their youth, testifying to extreme fragility.

At the two extremes of the scale, we have chosen to present two deltas that are almost intact, since they are located at the mouths of watersheds that are affected little by human actions, namely the Mackenzie and the Lena. We have then understood the contrast with the deltas of Southeast Asia that have undergone intense urban development in recent decades, accompanied by the characteristic impacts of this type of growth (extractions, water pumping, etc.). Thanks to these case studies, we have been able to summarize the effects suffered by the deltas. The fact that they are recent does not allow us to jump to conclusions about their intensity and their significant role in the degradation of deltaic environments; they allow us to understand the evolution of large deltas around the world.

Chapter 2 will present deltas that were enhanced by ancient societies and that simultaneously experienced an environmental history that reveals continental hydrosedimentary fluctuations.

2

Old Societies and Deltaic Crises

Deltas are zones constructed by sedimentary materials of continental origin that are partly redistributed by the sea. They are 6,000 years old at the most, even less in their distal parts, which goes a long way towards explaining their fragility in the face of oceanic forces. Paleo-environmental approaches have revealed that there has been a series of constructive and destructive phases throughout the history of deltas. This fluctuation is due to the variation in water and sediment inputs of continental origin over the course of millennia and centuries (for reasons that combine natural and societal causes), as well as the new and brutal changes they have undergone in the contemporary era. Chapters 4 and 5 of Volume 1 referred to several factors of human origin in order to explain flow stoppages, in particular the interception of flows by reservoirs. Recent profound disturbances can even go as far as causing deltas to disappear. Over the last few decades, we have observed what can be described, at best, as an increase in risks, and, at worst, as tangible signs of the sea returning to the areas that had been conquered by the combined factors of climate, water flows and transport of sediment. The situation is at a tipping point that puts remarkable water-based civilizations in danger.

2.1. Some vulnerable deltas in the Holocene during the long and medium terms

We will look at a series of contrasting situations as they have been expressed in the medium and long terms for the last six millennia. The first one is the Nile Delta, the future existence of which is highly threatened by the impact of the Aswan Dam; it underwent a natural crisis in its prehistoric past, under conditions that also involved factors common to the current situation; the second one is the Huang-He Delta, where the weight of the sedimentary heritage is so strong that it could somewhat immunize the delta for the centuries of shortages that are to come thanks to the stock of alluvial materials created in its valley downstream from the mountains; lesser reactions have been observed in the Mediterranean deltas (Rhône Delta, Ebro Delta, Po Delta and the Danube Delta), where variations at the scale of the basin have been accommodated through other factors, in particular the relatively small solid load of the Holocene rivers, even during the periods of erosive crisis.

2.1.1. *The Nile Delta, a condensed version of the history of the African climate*

The Nile Delta has undergone the physical processes that are classic for this type of environment. Worldwide deceleration of the rise in sea levels around 6,500–5,500 BC was the primary cause of accumulation of fluvial silts on the alluvial sand in the former riverbed, rediscovered in its buried location. Soil fertility relied on this silt; hence, the former's organic content was enriched to the level of sedimentary input that came from its basin and from the organic production that is specific to a permanently wet area. The first human occupations in the predynastic era (7,000–5,100 years BP), based on agriculture and livestock farming, were a logical consequence of the characteristics of this amphibious zone that had developed in a desert-like regional environment [STA 93a].

The rise in sea levels is not sufficient to explain everything about the characteristics of the Nile Delta. Recent research has shown that there have been significant and complex responses to climate change that the river basin itself has undergone in the last eight millennia [MAR 12, MAR 13]. On the basis of surveys and radiocarbon dates, we now know that aggradation of the deltaic plain occurred at an average speed of 17 cm/century, but with a gradual decline between 7,700 cal BP and 1,300 cal BP. The high values at the beginning of the Holocene can be linked to the abundant rains of the summer monsoon that fell in the upper part of the Nile basin in response to a greater exposure to sunlight. Consequently, sediment transport was greater at the time than it has been in the most recent millennia. These inputs largely exceeded the destructive forces in the delta, namely its subsidence and the rise in level of the Mediterranean Sea, with one adding to the other to lower its surface.

Other than aggradation of the delta above the water level, a period of drying-out is thought to have contributed to reduction of the Nile floods. Around 5,000 cal BP, the intensity of the summer monsoon reduced, with the intertropical convergence zone retreating to around 15°N, and North Africa entering a long dry period. Around 4,000 cal BP, sediment input no longer compensated for the effects of the rise in sea levels, breaches in dune bars and saline intrusions in a curved delta that had become wave-dominated; the environmental conditions of the upper basin meant that the Blue Nile, the main tributary, carried less water and less sedimentary material to the Nile. Rarefication of flooding of the Nile and of its branches was favorable to human and agricultural colonization of the delta during the period of aridification of the Sahara. The wave of occupation was based on marsh drainage and simplification of the network of distributary branches. In the context of complex interacting factors, these changes, a combination of natural deciding factors and the first human actions, encouraged subsidence of Holocene deposits by the reduction of the water table and the oxidation of organic sediments. A slow, unfavorable evolution was masked by the original fertility of the ground. The increasing aridity and reduction in input for sedimentation and fertilization was a cause of famine around 4,200–4,050 cal BP. The crisis at the start of the 4th millennium BC thus revealed the intrinsic fragility of the delta [MAR 13].

The Aswan Dam has a major role of responsibility for the current dynamic of the Nile Delta, but the work of geoarchaeologists has the great merit of contributing to the long-lasting environmental history. The latter reveals the fragility of a deltaic environment that is subject to the reduction in the flows of both water and solid matter in a context of increasing aridity. Famine arrived within a few centuries.

2.1.2. *The lower Huang-He and its delta: a Holocene metamorphosis under anthropological control*

Over the last 8,000–9,000 years, the downstream course of the river and that of the alluvial cone of the Huang-He has increased in elevation by 20–30 m, and the slope has remained almost the same. As we saw earlier, the river is supplied by a load made up of fine sands and silts that come mainly from the Loess plateau. Aggradation of the watercourse that generates this cone may occur several hundred kilometers upstream from the delta, since it begins in the downstream course of the river, upon exiting the mountainous region. Over the course of the history of the river in the Holocene epoch, aggradation has caused at least 10 diffluences* in one part or another of the Shandong peninsula; this figure is lower than the true figure and corresponds to the construction of 10 deltaic "super-lobes" located by geologists in the Bohai Sea and the Yellow Sea. The latter is the one produced by the 1855 diffluence event.

2.1.2.1. *At the historical origins of accelerated erosion on the Loess Plateau*

Work based on ancient scripts puts a date on the beginning of this accelerated erosion as approximately 2,000 years ago. During the last millennium, and especially during the last two centuries and before the construction of reservoirs, the sedimentary flux of the Huang-He that arrived in the delta increased from 0.1 to 1–1.2 billion tons/year due to clearing and the increase in agriculture on the Loess Plateau [MIL 87]. However, calculations made of the volume of sedimentary deposits in the delta apply a much more recent timeline to the erosion crisis, describing it as "abrupt" and attributing an age of only one millennium to it.

While the erosion crisis is younger, densification of occupancy of the Loess Plateau does nevertheless go back 2,000 years. The population of the plateau increased from 700,000 inhabitants under the Han dynasty (206 BC to 220 AD) to 60–70 million under the Tang dynasty (618–907 AD). Before 2,500 years BP, the Huang-He was nicknamed *Da He*, the "Great River"; then under the Han, it was the "Turbid River", before becoming the "Yellow River" under the Tang, a surprising semantic shortcut to illustrate the perception of a river through its relationship with sedimentary transport [SAI 01]. In agreement with the principles mentioned in the previous chapter, the time delay is due to the trapping of sediments in the alluvial plain at the beginning of the erosion crisis, which are therefore not as abrupt at the scale of the fluvial system as recent works in the region of the delta itself would suggest.

2.1.2.2. *Elevation of the lower Huang-He onto its alluvial materials*

Was the river capable of accommodating this considerable aggradation in its journey across the plain? The downstream part of the Huang-He is famous for its course that "hangs" above the alluvial plain, with dykes or "artificial levees" that have contained it, certainly to a greater or lesser effect. Their construction began during the Warring States Period (475–221 BC), probably around 350 BC, and reduced both meandering of the river and uniform spreading of its alluvium over its alluvial fan. However, it is obvious that the first designers were very cautious about containing the river within too narrow a course. In the first centuries of protection, the interdyke zone was required to be wide, up to 25 km, so that the inundated land could benefit from the fertile silt, and breaches were accepted. Subsequently, alluvial build-up was restricted to the zones between the dykes. It is interesting to note that Chinese literature (translated into English) describes the "floodable plain" of the Huang-He as only the area between the levees, although the external area can also be considered as floodable during rupture of the dykes. It is true that the zone located between the dykes had a width varying between 3 and 20 km, and contained 900 towns and villages populated with more than a million inhabitants in 1990 [SHU 93].

On the lower course and the deposition fan that has formed on the lower Huang-He, and when all information sources are combined (historical maps, annuals and chronicles, archaeology, paleoenvironmental studies), historical records show 1,593 levee breaches and 26 major changes of course in 2,550 years, which gives an average of two ruptures every three years (Figure 2.1). More than 300 paleochannels have been tracked in the field in the province of Hubei alone. These changes of course are accompanied by human disasters, with, for example, 310,000 deaths at Kaifeng in September 1642. In 1938, the advance of the Japanese army was slowed down by voluntary destruction of the levee on the right bank at Huayuankou and by flooding an area of 45,000 km^2, causing floodwaters from the Huang-He to reach the Huai He valley. In the provinces of Henan and Anhui to the south of the river and reaching all the way to the lower course of the Yangtze, 890,000 inhabitants lived in the area that was flooded [SHU 93].

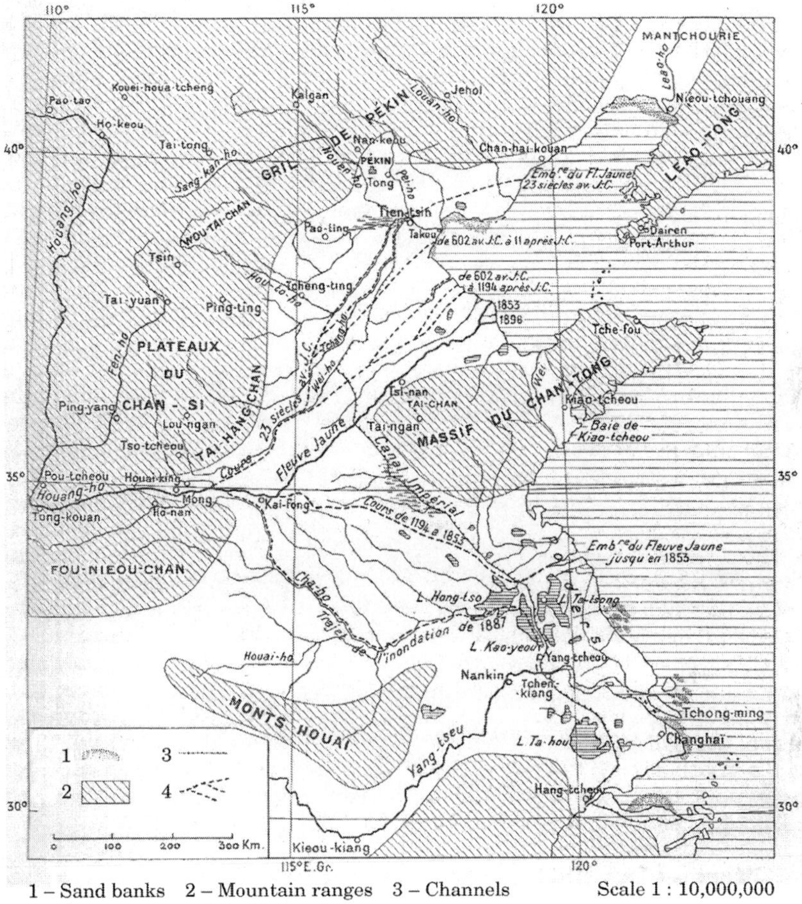

1 – Sand banks 2 – Mountain ranges 3 – Channels Scale 1 : 10,000,000
4 – Former course of the Yellow River

Figure 2.1. *Hydrography of the great plain of Northern China. The map of the successive courses to the north of the Shandong (Chan-Tong) massif, in particular the course resulting from the 1853 diffluence, and the course followed to the south of the massif between 1194 and 1853 [SIO 28]*

2.1.2.3. *Modern dyke construction works*

After 1570, the hydraulic policy of Northern China was modified to systematically raise the levees in a narrowed bed. In this era, the aggradation of the riverbed was 25 mm/year. This was also logically the period of the strongest and most dangerous avulsions, practically every time a breach appeared [CHE 12]. Under the Ming dynasty, from 1540 onwards, the renowned engineer Pan Jixun developed a theory and put it into practice: since the depth of water between the dykes was higher than in a natural configuration, the floods had a greater sedimentary transport capacity, a capacity to scour out materials, which would ultimately reduce sedimentation. The factor that Pan Jixun had not foreseen was that the increased length of the course of the river through its delta, augmented by the policy that he had put in place, reduced its slope and therefore canceled out his efforts. Deposition in the delta produced a regressive accumulation in such a way that aggradation of the channel increased. The failure of Pan Jixun's strategy was confirmed in 1855 when the Huang-He changed course following rupture of its dykes.

2.1.2.4. *Contemporary construction works*

Since then, the levees have been raised by 2–6 m and reinforced effectively, three times since 1947, over a length of about 800 km. They can contain a maximum flow rate of 28,000 m^3/s, which was the flow rate of the 1958 flood, the return period of which is estimated to be 60 years. Since this date, no rupture has occurred, but the rate of accumulation has now reached 8 cm/year, a record in the history of the Huang-He. The capacity of the dyked channel is considered insufficient despite the efforts made, and a future catastrophe is probable, all the more so because this is a seismic region, and sometimes the levees and infrastructures are cracked, as in an accident that occurred in 1983 [XU 98].

Recent digital modeling has to a certain extent confirmed the negative impact that the levees constructed along the Huang-He have had throughout history [CHE 15]. It confirms the well-known principle that raising dykes increases their relative height, their risk of rupture and therefore the risk of avulsion. This policy of fixing the riverbed in place turned out to be very costly for the state since, between 1670 and 1839, 4% of China's expenditure was allocated to the maintenance of the levees, in particular on the largest breaches. This policy has to some extent been trapped by the initial, now-distant decisions. Chinese researchers admit that the policy implemented since 1960, which diversifies the methods of treatment by attacking the root of the problem (erosion of loess materials) and choosing to construct dams on the median course of the river, has greatly reduced the frequency of problems inherited from a distant past.

2.1.2.5. *A delta that is still river-dominated despite the decrease in inputs from the Bohai Sea*

It is difficult to find a delta on Earth that is less river-dominated than the Huang-He; its unique form bears witness to this. The apex of the delta-fan is at Kaifeng, at an altitude of about 80 m and a distance of 650 km from the Bohai Sea, and taking into account the slope of the channel, the influence of the tide in the lower valley can only be felt over 50 km compared to 600–650 km for the Yangtze (at the Datong station). If we again consider the figure 1.6 Gt/year that passes at Sanmexia, of this total amount, 0.4 Gt are deposited in the channel of the lower course, about 0.8 Gt in the mouth zone and still 0.4 Gt in the shallow waters of the Bohai Bay or Sea. The fact that the concentration still approaches 1,000 kg/m^3 in the lower part of the river explains in part the intensity of the deposits. Estimates for the period 1951–2000 give an average annual accumulation of 0.11 billion m^3 and a cumulated total of 5.4 Gt.

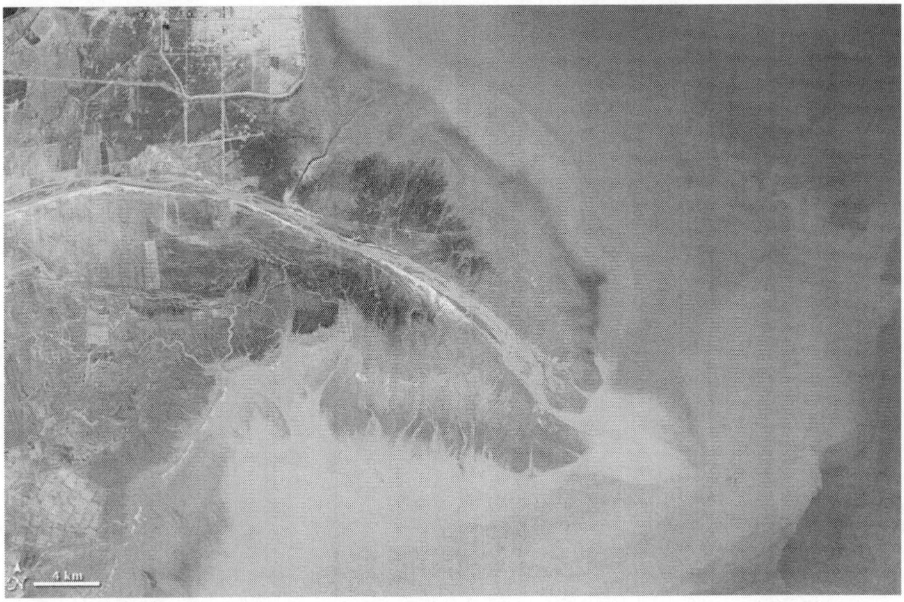

Figure 2.2. *Image of the Huang-He Delta taken on February 13, 1989 during the dry season. The small peninsula undergoing construction in the Bohai Sea is oriented southeast in an environment that has still seen very little development. (Source: NASA Earth Observatory). For a color version of this figure, see www.iste.co.uk/bravard/sedimentary2.zip*

The shapes of the old coastline, materialized by bars made up of shells ("cheniers"), reveal that it was influenced by wave dominance, which is due to a moderate level of continental input; the chenier dated 8,000 years BP is set back, at a distance of 50–100 km from the current coast. The penultimate lobe dates back to the period 1128–1855, and the most recent has been forming since 1855, which is the date of the most recent diffluence. The current lobe has an extended form in the shape of a crow's foot, which demonstrates the influence of the river, the small amplitude of the tides (the Bohai Sea being microtidal*) as well as the low average height of the waves, where the latter are, however, dominant in terms of shaping the details of the coastline [SAI 01].

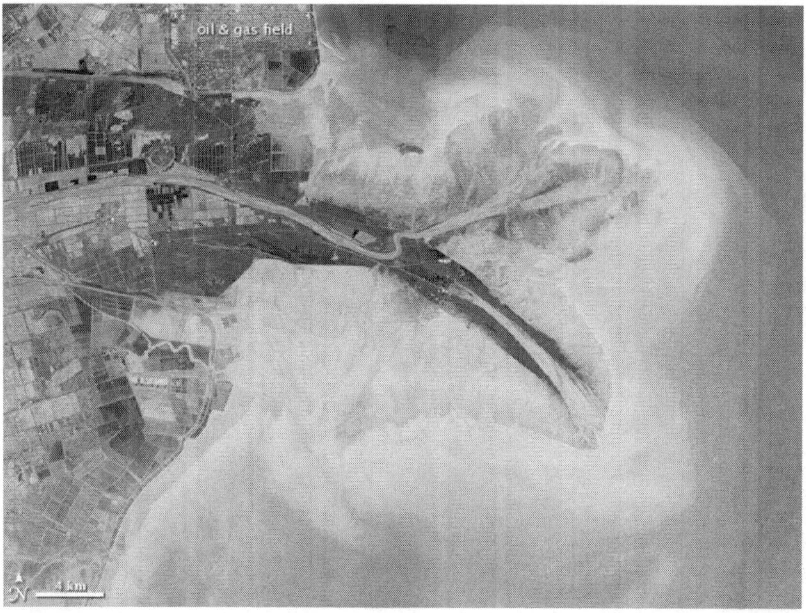

Figure 2.3. *An image of the Huang-He Delta taken on June 20, 2009. The peninsula under construction has had two lobes since a channel was opened in 1996 (and the old closed channel) to encourage the growth of the northern part of the growing peninsula. Establishment of exploitation on the Shengli oil and gas field (the second in China), protected by dykes from the sea, and development of irrigated marginal areas and subsequently of aquaculture, have caused protection measures to be put in place in an environment marked not only by the drying-up of liquid and solid flow rates, but also by aggradation of the riverbed that is raised above the plain. (Source: NASA Earth Observatory). For a color version of this figure, see www.iste.co.uk/bravard/sedimentary2.zip*

The Huang-He Delta is completely atypical, as its sedimentary history has explained to us. On the one hand, the delta is subject to river dominance, and on the other hand, and not unrelated, its terrestrial part (the "upper delta" according to the general nomenclature) is mixed with the river's alluvial plain up to more than 600 km from the sea. The area for sedimentary accommodation in periods of excess and shortage is considerable, and the variations recorded in the distal part bear the stamp of this. Centuries of sedimentary shortages would be required to modify the general configuration of the Huang-He at its mouth (Figures 2.2 and 2.3). The Huang-He Delta appears to have been somewhat immunized by the sedimentary history of the fluvial system that controls it.

2.1.3. *The Rhône Delta during the Holocene: fluvial branches and the coastline record the history of its climate and society*

The Camargue and the neighboring alluvial plains of the Grand Rhône and the Petit Rhône, spread over an area of 1,740 km^2, constitute one of the large Mediterranean deltas. The level of the sea, very closely related to that of the oceans, rose very quickly up to around 8,000 years BP. It was at its lowest level at the time of the peak of the most recent glaciation, around 18,000 years BP. The relative rise in waters slowed down from 7,000 years BP onwards. It reached a stable level at – 2 m below the current 0 level between 6,300 and 4,100 years BP, and then it began again around 2,300–2,200 years BP, at an average speed of the order of 2 mm/year, since this rise in waters is partly due to the settlement of sediments [ARN 98, ARN 02, PRO 02]. As with the other deltas around the globe, the Camargue is therefore a young developing zone when the impacts of the industrial era came into play. It has therefore been all the more sensitive to the imbalances that have been imposed on it.

Sedimentary construction of the Camargue delta has followed a specific process, which has combined a continental trajectory and a dynamic that is unique to the deltaic zone. In the history of the Holocene dynamic of the Rhône basin, now quite well known, combined phases of hydrosedimentary crises and remissions have been observed; the crisis phases mostly have a double climatic and human causality (Volume 1, Chapter 1). In the delta, which, as a formation located at the basin exit, records the fluctuations, fossil landforms show the internal variability of the basin. During 10,300–8,300 years BP, the sea swamped the land areas that had remained above sea level during the phase of rising sea levels, which led to the deposition of marine sediments. The coastline then ceased to progress on land because sediments of the Rhône constructed the fluvial delta faster than the rising sea level. The resulting deltaic plain progressed by the progradation of lobes that are probably fed by the Grand Rhône and the Petit Rhône, as well as by several fluvial arms that are currently abandoned: the Rhône de Saint-Ferreol flowed between 6,000 years BP and the Middle Ages (11th–13th Centuries); the Ulmet arm appeared around 4,500 years BP, contracted at the time of the Romans and was closed off during the 15th Century. The coast regulates itself between the lobes by prograding into the sea with the presence of parallel dune bars that mark out a convex lobe; they are spread over several kilometers to the west of the former Rhône de Saint-Ferreol branch that fed them with sand. One of these bars, dated from the 2nd–1st Century BC, was located 3 km off the coast of Saintes-Maries-de-la-Mer by an alignment of shipwrecks on the swamped pre-coastal bars, a sign of regression of the coastline since this period.

Infilling of old river branches and sedimentation on their edges provide valuable information about the fluctuations in their hydrology and about the sedimentary volumes that come from the watershed as time passes. A low level of activity was observed during the period that corresponds to the La Tène period (from the

5th Century BC to the end of the 2nd Century BC). The wet period with very large floods that took place in the 1st Century BC and the 1st Century AD can easily be identified in the channels where sand banks are formed, with these channels aggrading and shifting laterally. This period gave way to a period of calm or hydrological "irregularity" in the 2nd to the 5th Century AD, with incision of the channel and a low water table, reduced deposition of fine sediments and formation of thick soils. The period of the 5th to 7th Century AD has seen a new hydrological crisis, with infill by coarser sediments before a new, long respite from the 8th to the 15th Century AD; this respite itself preceded the last crisis of the Little Ice Age. While the phases of hydrological calm encouraged the colonization and movement of humans on alluvial bulges, those of the periods of hydrological crisis or at least relative irregularity seem to have been characterized by relative abandonment [LAN 04].

2.2. The Rhine and the Meuse Deltas: from complete control of fluvial and marine waters to attempts at restoration to a natural state

The name given to the Netherlands suits it well: 9 million of its inhabitants live below sea level and 70% of the current GDP (gross domestic product) comes from areas located below sea level [MUL 11]. The delta of the Rhine, the Meuse and the Scheldt is the epitome of a delta that has gradually become entirely developed through the centuries. These questions echo those in the first chapter of Volume 1 of this book, in which the hydrographic network and the first fluvial developments have been presented; the problem with the Rhine echoes the question about the excessive build-up of alluvial load in the mountainous regions of Europe, as well as that of the means to remedy this in order to ensure security and maintain the use of the river. After a rapid presentation of how the deltaic surface is constituted, we will see that control of waters in the Netherlands occurred in two phases: on the one hand, protection of floodplains with respect to fluvial flooding; on the other hand, protection of the deltaic plain against the North Sea.

2.2.1. *The fight against fluvial floods*

One flood of the Rhine, in January 1809, caused serious flooding. It is a good illustration of the fluvial threat that presents itself repeatedly to the Netherlands [WIK 18a]. Rupture of the dykes flooded the Betuwe, a small region located between the Waal and the Nederrijn-Lek rivers further to the north, downstream of Arnhem, as well as part of the Biesboch itself which is affected by the Meuse floods. The seriousness of the event was due to the ice jam of the deltaic branches on which the fluvial flood arrived, produced by the oceanic rain that fell on the upper part of the basin. The 1809 flood is due to the conjunction of the effects of cold and rain on the upper basin, but the fluvial risk has always been present in various forms, making the control of the branches unavoidable.

2.2.1.1. *Technical control of the branches of the Rhine and the Meuse*

The Rhine (1,320 km long) drains a basin with an area of 165,000 km^2. Upon its arrival at the top of the delta, the Rhine has a flow of 2,200 m^3/s averaged over one year, with a maximum in February and a dry spell in October. Concerning the Meuse (950 km long), it drains a basin of 36,000 km^2 and has a flow of 357 m^3/s averaged over one year at the top of the delta. Holocene deposits of the delta lie on top of gravels laid down during the last glacial period (the *Weichselien*) by braided channels with a strong slope flowing towards the North Sea. The facies of Holocene deposits of the undeveloped deltaic plain were made up of:

1) silts and clays deposited by floods and autochthonous organic formations (peats) in the basins; today, measurements made of these formations reveal a thickness of 1 m at the German border and gain in thickness further to the northwest, up to 20 m;

2) sands deposited in lateral banks by the watercourses and shaped by their shifting movements, as well as deposited downstream from the breaches opened in the natural levees at the time of floods.

Clay was present in the infilling of abandoned meanders. This architecture ceased to exist as dyking was carried out [HES 02]. The coastline features beaches and sandy dunes that were reworked during the Little Ice Age.

Double dyking of the deltaic branches of the Rhine and the Meuse was in place between 1050 and 1350, with the aim of protecting the plain from flooding, to claim agricultural terrain and secure a rapidly growing population by eliminating the fluvial meandering and changes of course. Construction works in the 18th and 19th Centuries normalized the situation, i.e. stabilized the distribution of flow rates and blocked in the geometry of dyked beds between powerful structures (refer to chapter 1):

– the high dykes (levees) of the winter riverbed (500–1,000 m wide), within which the narrow alluvial bands of dyked major beds develop (total surface area 380 km^2); these are generally flooded from December to April for flow rates over 3,500–5,000 m^3/s depending on the section;

– the low dykes that contain the summer riverbed (at low flow) and the small floods between May and October.

Behind the hydraulic structures that contain the fluvial waters, the deltaic plain is a polder divided up into a large number of small units that are equipped to evacuate the water that has accumulated from precipitations and any organized or catastrophic overflows. This is another variable that the Netherlands must manage: that is, slow subsidence of the delta, due less to settlement of Holocene sediments than to the undesirable effects of underground contemporary water pumping. Their impact contributes to the topographical differentiation between the fluvial bands and the adjacent basins.

2.2.1.2. *Flood management and management of fluvial flooding in the 18th and 19th Centuries*

Medieval construction works obtained the effect (that they had sought) of containing the flow rate of the flood of deltaic branches between dykes. The increase in energy due to narrowing of the main channel caused *in situ* incision, whereas sediments carried down from upstream were deposited between the low dykes and the high dykes over an area known as the “dyked alluvial plain”. The flood water circulated higher than the flood basins, as a result of which they were transformed into polders. Any rupture of high dykes was a real danger because the basins were inundated under a water layer of a thickness that increased downstream; near the breaches, the speed of water posed the danger. The situation was even more serious because the sedimentation between the dykes was reducing the storage capacity of the dyked alluvial plain gradually. The water levels have been recorded daily at several measurement stations since 1770, which create a database that is unique on Earth. Certainly, the very good knowledge of the history of the developments, sedimentation, breaches and flooding has given the Netherlands a high level of expertise in terms of management, habitat localization and means of evacuation of the population since the 17th Century.

2.2.1.3. *Correction of the sediment deficit of the deltaic zone*

The cumulative sediment deficit of the delta in the Netherlands, which is intersected by dykes, is estimated to be 136 ± 67 Mm^3/year from now to the year 2100, taking into account a rise in sea level of 35–85 cm; this deficit has been estimated with respect to a deltaic plain that is rising along with the river due to free sedimentation, which is in fact no longer the case: 85% of the deficit is due to the lack of compensation for the effect of the rise in sea levels and 15% to drainage that will lead to compaction and decomposition of the peat. The volume that is not used by sedimentation will reach the levels of 13.3 km^3 per year, which weighs on the resilience of the delta. The policy is to leave a free choice of ground space attribution to the inhabitants, relying on the existing protection, but at the risk of dispersing efforts too widely in the event of a dyke breach. With this in mind, a solution to this problem would be to artificially raise the level of terrain that is dedicated to infrastructure and open to construction; later on, the policy was to abandon the shallows as wet areas and move agriculture up to aggraded land. A program of systematic infilling of the area with marine sand at a rate of 12 Mm^3/year would resolve half of the problem within a century [MEU 07].

2.2.1.4. *Anthropological impacts and sediment deficit of the Rhine*

The bed of the Rhine channel is formed from gravel at the head of the delta but of sand downstream. Transport of sand and gravel at the border between Germany and Holland was 0.66 Mt ± 40% per year between 1991 and 2010, which then ceased downstream before arriving in the North Sea. Yet this type of sediment was absent from the channel before the medieval period during which the first great series of works took place. Therefore, it is necessary to consider that thickening of the bedload is due to a combination of impacts:

– medieval dykes and intersections of modern meanders have narrowed the bed of the Waal, increased the height of water during floods and the shearing forces along the bed, thereby allowing large particles to be dragged downstream;

– the works carried out after the 1740 floods have been followed by installation of groynes with the aim of accelerating the evacuation of floodwaters and then of improving the conditions for navigation. These structures have extended and accentuated the previous impact. The bed has been incised in places where it has not been dredged. The only possibility for adjustment that was left for the Rhine channel was the increase in size of the bed particles that form an armor and block the incision.

The solution selected by the managers of the Rhine to stop, or compensate, for the channel incision has been to reload it with small gravel in areas where there is a deficit of this, without being able to modify the hydraulic parameters. The origin of the gravel is divided into three approximately equal thirds: descent of the bedload from upstream, erosion of the bed and reloading (which inhibits incision). The question, without being crucial, deserves to be studied in order to correctly understand the sedimentary budget of the delta. For example, erosion of the bed produces sand that is deposited in the dyked alluvial plain and participates in clogging of the plains by sediments. The strategies of the managers will depend on the expected researched inputs [FRI 15].

2.2.2. Hydraulic works and environmental objectives in the dyked zone

The floods of the Meuse and the Rhine that occurred in 1993, 1994 and 1998 have been responsible for the creation of breaches, the flooding of polders and the evacuation of 250,000 people in the Netherlands. These floods demonstrated that sedimentation between the high dykes had exceeded the tolerated level. The 1997 plan prepared by the RIVM[1] was primarily destined to fight against fluvial flooding;

1 RIVM: Rijksinstituut voor Volksgezondheid en Milieu. It is the Netherlands National Institute for Public Health and the Environment.

this question has also become a major concern for the International Commission for the Protection of the Rhine (IPCR) and for others that have occurred in Europe (e.g. the Vltava and the Elba).

The IPCR Flood Defense Action Plan (1998) duly notes the responsibility of the countries in the Rhine basin in the increasingly serious nature of flood peaks all along the Rhine and particularly in the Netherlands. Water management for the delta is also carried out at the scale of the basin. With the aim of reducing the height of flooding in the delta by 70 cm between now and 2020, measurements have been made to retain the water in the sections of the alluvial valleys and to increase their storage capacity. The basic principle, selected at the European scale, is to manage the floods in other ways by giving space back to rivers, by reducing the amplitude of low and high waters that have become too high due to past development, and by slowing down the downstream progression of waves of flooding by directing peak flooding into storage reservoirs. This is also the philosophy applied by the authorities in the Netherlands, who use the same measurements to reduce the height of a reference flow rate of 16,000 m^3/s, which has become "the design flood" (flood with a return period that has increased to 1,250 years). Two opposing methods exist: the official position, which was to increase the height of the existing dykes and to accelerate the exit of waters towards the sea, and the environmentalist position, which advocated water storage according to the European model favored at the time. On this basis, two types of measurements have been set up. On the one hand, structural measurements have been set up to compensate for the negative impacts on the flow; thus, it was decided to push back sections of dykes, to remove all obstacles to flow in the dyked plain, to clear the surface of deposits, or to extract the sand and gravel along the lower Meuse[2] to recreate a storage volume. This policy, applied in the area located between the winter dykes, makes even more sense (and obligation to succeed) because the Action Plan has designated polders to be sacrificed in the case of exceptional floods (nicknamed "calamity polders"), even though they had been populated and economically developed under completely legal conditions; the conflicts in these zones are quite similar to those that have led to the abandonment of the flood expansion areas in the lower and middle Rhône valley.

On the other hand, the RIVM also has environmental targets, since it makes provision for the reconstruction of the natural landscape of over 18,000 ha in the dyked alluvial plain with conversion of agricultural terrain into a landscape of dynamic channels, marshes, lawns and soft and hardwood forests that have been restored to increase the roughness of the dyked alluvial plain in order to slow down its currents and to store water more effectively. This ambitious project is based on

2 The project *Living Border Meuse*, implemented from 2000, brings together Belgium and the Netherlands on the border traced by the Meuse. The objective is to restore fluvial dynamics by depositing riprap over a distance of several kilometers.

the concept of "habitat as a model for organizations", meaning that if we are capable of recreating suitable habitats, with spatial and temporal dynamics, it will be possible to recreate a healthy and operational natural environment. The project takes "site engineering" (a different concept) into account, according to which the species will settle in by themselves and will prosper with a minimum of interaction with human societies.

2.2.3. *What kind of compatibility or synergy takes place between fluvial restoration and protection against flooding?*

The first reservations about these nature restoration projects came from hydraulic engineers. They estimate that the factors responsible for sedimentation are always active, and that the works carried out can only accentuate the deposition trend. Localized removal of groynes will certainly allow lateral erosion of the Waal river channel, and will therefore increase its width and reduce its energy, but these processes will encourage sedimentation in other places. In other words, the new situation will bring the river closer to the conditions that existed before normalization, and it is highly probable that the sedimentation will begin again in the IJssel and the Nederrijn.

However, when we admit that this obstacle can be overcome, is it reasonable to hope to obtain restoration of natural landscapes, as many ecologists and managers hope to see [NIE 01]? It has been noted that the possibility of returning to natural processes in rivers in the Netherlands is very limited due to the regulation of the flows by levees and the geometry imposed by the dykes. For navigation requirements, which are considerable, the channels have been transformed or adjusted in such a way that they are deeper and less wide, but subject to restrictions. In view of the flow rate carried by the Waal, the natural average width of the channel should be 500 m, not 170 m as it is today, i.e. three times greater. As a result, any return to spontaneous migration of meanders is excluded, because it assumes occupation of more space within a large migration belt. According to engineers, it is doubtful whether the areas that are realistically possible to dedicate to the improvement of flood hydraulics in Europe can combat the amplitude of the problems:

> "We should realise that the deteriorated large rivers, as we know them nowadays, leave only little space for rehabilitation measures, without violating the basic rules of safety, unless very draconic measures should be taken" [NIE 01].

Similarly, recreating a vegetated landscape in the dyked alluvial plain, such as the one prevalent around 2,700 years BP (drawing solid lessons from pollen studies) or in the Middle Ages, is impossible, because it was a case of hardwood forests with a few clearings cut by farmers. It is impossible to recreate the natural conditions for the physical basis that leads to the natural succession of vegetation, except for creating conflicts with other users of this area.

2.2.4. *Defense of the Netherlands against the sea*

Whatever the efforts made on the branches of the Rhine and the Meuse, the risk from the North Sea has, however, been considered to be greater than the risk of fluvial origin for around 60 years. The funnel shape of the North Sea makes the coasts very sensitive to storms coming from the northwest and north.

The typical example is the Sainte-Catherine storm, which occurred in the night of 31 January 1953. An exceptional storm surge by the sea, due to a force-10 storm from the northwest, which occurred at the same time as a spring tide, drowned the estuarian mouths under about 4 m of water, blocking the fluvial flow. The mediocre maintenance of the dykes since the economic crisis and during the years of war and the lack of financial means after the war, with priority given to the reconstruction of the country, partly explain the serious nature of the damage in the 117 flooded villages. The disaster resulted in the deaths of more than 250 people and 30,000 animals; it destroyed 10,000 dwellings and flooded 9% of farms in the Netherlands (Figure 2.4). The trauma caused by the Sainte-Catherine floods led directly to the design and implementation of the Delta Plan in the Netherlands, one of the seven wonders of the modern world according to the American Society of Civil Engineers, as well as to that of the Thames Barrier in England [WIK 18b].

2.2.4.1. *The Delta Plan (1956–1986)*

In mid-February 1953, the Delta Commission was installed by the Minister for Transports and Water Management. The decision to construct powerful dams across the mouth of the deltaic branches rather than reinforcing old marine dykes that protect the banks of the estuaries was made in 1954; the principle was to reuse the model of the Brielse Maas[3] dam that was completed in 1950. Only the Western Scheldt (the southernmost arm of the Sea of Zeeland) was not dammed, because it leads to the port of Antwerp.

3 The Brielse Maas is the Meuse de Brielle, the former estuary of the Meuse that has been filled with sand and doubled up by a navigation channel dug in 1872.

Figure 2.4. *Map of the regions of the Netherlands that were flooded in 1953. (Source: Diercke Schulatlas, Verlag Westermann, CC BY–SA 2.5). For a color version of this figure, see www.iste.co.uk/bravard/sedimentary2.zip*

The Netherlands Delta Plan was carried out in several phases between 1957 (the date of its official adoption) and 1976 (the date of its inauguration); it was then restarted and partially completed. The construction designs used have varied over time, which is not at all surprising since the overall construction was carried out over a period of around more than 50 years. The most significant characteristic of the Delta Plan, explained in part by its very duration, is the permanent adaptation of the policy followed by the deciders, which fits with the evolution of designs in terms of sustainable development of the estuarian environment.

The first phase (1957–1973) had the sole objective of preventing sea water from entering or invading by means of the estuary barrier, and creating lakes of fresh water. From north to south, it includes the following main elements:

– the dam of the Harengvliet sea arm (1958–1971), which receives the Hollands Diep that gathers together the waters of the Meuse and the Rhine (with the sea arm existing since the 1216 storm);

– the Volkerak dam (1957–1969) on the south branch of the Hollands Diep;

– the Greveling dam (1958–1973) that partly closes the mouth of the Eastern Scheldt (5 km of 9).

The structures of the first phase were severely criticized because they destroyed the interactions between the coastal waters and interior waters, did not allow for sustainable development and finally increased the vulnerability of the zone to subsequent disasters.

The second phase (1974–1986, then 1991–1997) perfected an additional plan set-up to take account of the protests that were aired following the completion of the Eastern Scheldt project. The main structure is the anti-storm barrier of the Oosterscheldekering that guarantees the safety of the population of Rotterdam against an event of a return period of 10,000 years (4,000 years in Zeeland); constructed between 1976 and 1986, it is open under normal circumstances, but closed when the tide and/or the storms exceed a certain level. This phase has introduced a new policy, which is protection of the landscape and the estuary environment, with reintroduction of salt water and defense of the marshes. This orientation, described as passive, has the inconvenience of requiring intensive and continuous care, because the natural capacity for adaptation was not restored, according to specialists in the Netherlands.

Total protection was guaranteed by a final structure, the floating gates of the Maeslantkering or the protection barrier for the Meuse region (1991–1997). This additional barrier, constructed as part of the *Europoortkering* project, worked well in 2007 and protects the port of Rotterdam more effectively (Figure 2.5).

The third phase, that of "geo-ecological constructive management", was based on "understanding the operational characteristics of environments that are taken as elements and integrated into the structure of the landscape"; it is a case of restoring gradients, for example by reintroducing the influence of tides behind the structures and by restoring the salt marshes that are capable of contributing to sustainable development of the delta and allowing aquaculture to prosper [VRI 96]. Today, the Delta Plan barrier is 25 km long and considerably shortens the primitive defenses, whose extended length was more than 700 km. In its natural state, the estuaries of the Rhine, the Meuse and the Schelde[4] were muddy environments that guaranteed purification of water arriving from the continent. Completion of the Delta Plan isolated the masses of estuarian water from, on the one hand, the sea, and on the other hand, the rivers, in such a way that the dynamic of aquatic fauna is no longer

4 The Schelde is named the Scheldt in the Netherlands, which has a length of 355 km, of which 160 km is tide-dominated. Its estuary is the Westerschelde, partly in Belgium and partly in the Netherlands.

monitored other than by chemical and biological parameters. This operation increased the vulnerability of the water masses to external disturbances and reduced their power footprint in the same way as, in general, their function as stabilizers. Moreover, considering the rise in recreational uses and the environmental value of these spaces, conflicts with navigation and fish farming production operations have increased in number [NIE 01].

Figure 2.5. *The Oosterscheldekering, or Eastern Scheldt Dam (1976–1986). It is 8 km long, constructed of concrete pillars in which sliding doors are positioned in grooves and are lifted when dangerous high tides and storms occur. (Source: Rijkswaterstaat, beeldbank.rws.nl). For a color version of this figure, see www.iste.co.uk/bravard/sedimentary2.zip*

2.2.4.2. *Climate change hurdle for rivers and the coastline*

Forecasts of the rise in sea levels predict a figure of +1.30 m by 2100, whereas on the continent, forecasts predict more serious fluvial flooding of oceanic origin.

Concerning the fluvial component, several collaborative studies on modeling the possible effects of climate change have been carried out since 1989 at the request of the International Commission for the Hydrology of the Rhine Basin. These studies predict a reduction in summer flow rates and an increase in the flow rates in winter, in anticipation of oceanization of the climate and a lesser importance attributed to snow (i.e. moving from a hybrid pluvial/nival regime towards a pluvial regime). An increase of 10–20% in winter precipitation is likely to significantly increase the risk of rupture of so-called "winter" dykes, without accounting for the effects of sedimentation on the dyked alluvial plain, which will be significant by 2050.

Information about a large number of variables is missing from the prediction. For example, the reaction to hydrological change in the construction processes of the channels and of the dyked plain as it will be transformed by the works remains unknown, in the same way as the behavior of the water table and the biodiversity of the various units of the deltaic plain as a whole [MID 01]. Another size issue will be correct management of the operational relations in the continuum* formed by the rehabilitated sections, then in the estuarian exits blocked by the Delta Plan and finally the North Sea, with the general objective being to accommodate greater sedimentary loads and larger deposits during larger floods.

Concerning the coastal component, after several millennia of rising sea levels, the interior continental plateau no longer provides sediments. The modern coast of the Netherlands has been fed by marine sediments carried to the platform during the Little Ice Age, and this stock is running out; a change that results in erosion of beaches and dunes. Moreover, continuation of the process of rising sea levels will maintain this dynamic. The management system in the Netherlands, initiated in 1990, relies on the conservation of coastal dunes. This is why the latter have been aggraded by 5 m and the tidal flats reinforced artificially with sand to the north of the Zeeland barrier. Initial operations involved extraction of marine sand on the platform to reinforce the coast. After adding materials directly to the beaches, practices evolved towards deposition of sand onto the tidal flats, which is then progressively carried off by the waves and deposition along the direction of long-shore drift, towards the southwest. The objective is to widen the beach and reduce the abrasive action of the waves and the risk of scouring at the foot of the dunes. As part of the initiative known as *De Zandmotor*, a volume of sand of 21.5 Mm^3 will be deposited on an area a little larger than 1 km^2, and no longer with a view to it being picked up naturally and redeposited on a 10 – 20 km section of coast over a period of 20 years. The foundations for modeling this mega-input have been established and will be continued, even though modeling has not produced solely positive results in the long term, along the coast; an important point is that the cost estimate is below that of structural solutions [BROW 16].

2.3. Contemporary imbalances in the Old World

The lack of sediments is the common denominator of all dynamic deltaic processes in old developed deltas.

2.3.1. *A delta with a reprieve: the Nile Delta*

The Nile Delta covers an area of 24,000 km^2. It had 25 million inhabitants in 1990, i.e. 38% of the population of Egypt; today it houses 39 million of them, including 4.5 million in Alexandria. The human density of the delta has therefore gone from 1,040 to 1,625 inhabitants/km^2 in 25 years (Figure 2.6).

The delta, which was almost abandoned when Bonaparte's expedition set out, was recolonized under the authority of Mohammad Ali and his successors, giving preference to large properties. The creation of an urban network, incision of channels to facilitate exchanges and perennial irrigation, as well as dam construction, are the large construction works that have led to the improvements that finally came into effect at the end of the 19th Century [FAN 06a]. The delta provided about 40% of the agricultural production of Egypt and 60% of its fishing resources. Yet passively, it has suffered multiple instances of damage to its existence. Drought and the lack of flooding of the Nile certainly protect it from submersion by fluvial flooding that would be the cause of slow sinking, but it shows signs of an insidious crisis that is sounding its death knell. The delta now has only two active branches, namely Rosetta (Rachid) and Damietta. To the east of the rocky headland of Alexandria, the coast continues for a linear distance of 240 km. Moving eastwards, we will successively see the Bay of Aboukir, the Rosetta promontory, the convex and concave coasts of Burullus and Gamasa, the Damietta promontory and finally Manzala Bay and Lagoon, before reaching Port Said. The coastal fringe includes low-lying and muddy land, partly overtaken by fish farms, sand bars and lagoons. Since subsidence processes and the rise in sea levels add up, the delta becomes more vulnerable in the face of threats to the sea, whereas sediments from the mountains of Ethiopia that have constructed it are stored in Lake Nasser. Fresh water is certainly the lifeblood of the Nile Valley and Delta, but the joint action of water and sediments that make a river no longer participate in maintaining the various equilibriums.

2.3.1.1. *Subsidence*

The delta is continuously sinking. It has been demonstrated above that the process of geological subsidence has been continuously active during the Holocene. While Alexandria is constructed on a carbonated ridge from the Pleistocene epoch, in a region of uplift, the Bay of Aboukir undergoes a significant subsidence of 5–7 mm/year; submersion placed Ancient Greek (Ptolemaic) sites under water depths of 2–5.5 m. The ancient city of Canopus has sunk by 8 m since its construction. Considering the presence of resistant rocks under the western part of the delta, subsidence is mainly due to the tectonics of the underlying substratum, obvious for seismic activity. Subsidence of the delta, related to compaction and the loss of water from Holocene sediments, is today between 1 and 5 mm/year depending on the sites; it is more sensible in the east than in the west, because the Holocene sequence is thicker in the east as deep tectonics has reduced the level of the Port Said region [STA 93b]. It is also somewhat due to the extraction of underground water.

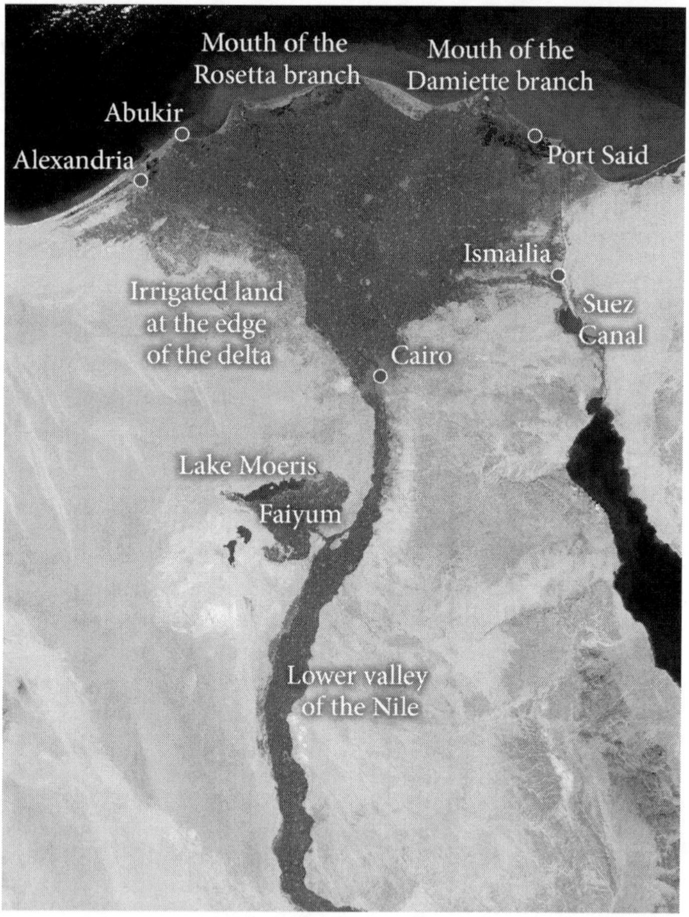

Figure 2.6. *The lower course of the Nile and its delta in their desert environment. Irrigated areas have been established on the edges of the river and the delta above the alluvial plain. The Faiyum depression with its lake fed by gravity from the Nile. (Source: USGS Nile Delta, NASA Visible Earth, Jeff Schmaltz). For a color version of this figure, see www.iste.co.uk/bravard/sedimentary2.zip*

2.3.1.2. *Relative rise of the sea level*

The rise in sea level is relative, since it is the sum of the eustatic rise, which is expressed at a global scale, and of the sinking that varies spatially. The relative rise recorded on the coast is modest since it is in the range of 1.8–4.9 mm/year, i.e. a value that frames the value attributed to the worldwide rise in sea levels and that incorporates subsidence; this is incompatible with the value of subsidence estimated above, which is greater (5–7 mm). One reason could be that subsidence has been estimated in the long term, and that the movements controlled by neotectonics (of which earthquakes are part) are active and today affect the deltaic space differently [FRI 03]. The areas that are most affected by the rise in water levels and by the risk of marine incursion during a storm are lagoons and low-lying land that cover 15% of the area of the delta. They are rich ecosystems with high fishing and tourism potential; moreover, a significant part of the internal zone has been converted into fish farms. The dune bars, still locally progressing, are altered by the construction of roads and houses, by agricultural colonization as well as by the extraction of heavy minerals, despite their obvious importance for protection of the delta.

The rise in sea levels, for which varied estimates have been made, could be between 25 and 100 cm from now to 2100 (probably more). Other than flooding of the low-lying areas, the Al-Salam irrigation canal that carries the water of the Damietta branch towards the plain of Tina (Sinai) is threatened where the accumulation of sea level rise and subsidence is the most marked. However, recent projections are in favor of a diversified scenario, quite far away from the very general visions of the World Bank that count on the assumption that an increase by 1 meter would cause flooding of a quarter of the delta and displacement of 10% of the population from now to the end of the century. Finer analyses estimate that long segments of coast would maintain a high position and a protective role; however, it is true that salinization of freshwater underground reservoirs is a real threat and already a sensitive issue for the land located behind it [FRI 10].

2.3.1.3. *Regression of the mouths*

Without continental input, deltaic mouths and their sandy coast undergo the destructive action of waves and currents, with the Nile being precisely the type of delta subject to these joint influences. The first work surrounding the question, carried out by comparing maps and satellite images, and published at the end of the 1970s, was about the change in the position of the promontories constructed by the Rosetta and Damietta branches and the behavior of intermediate coastal cells. While the promontory of the Rosetta branch progressed by 3.5 km towards the open sea between 1800 and 1909, it regressed by 53–58 m/year between 1909 and 1988 (especially between 1970 and 1988). It is due to the effects of waves and long-shore drift activated by a prevailing wind from the northwest/north-northwest that the sands of the Rosetta promontory have benefitted the eastern bay. The same observation is made on the Damietta promontory, but with a longitudinal retreat of 2.6 km between 1909 and 1983 (i.e. 18–28 m/year). The east flank of the headland regressed the fastest, with the formation of an arrow progressing towards the southeast and the displacement of about 1 km^3 of sand. Various studies have converged in their estimates that the trend inversion must be attributed to hydro-sedimentary fluctuations of the Nile under climatic control; the natural flows were greater at the end of the 19th Century due to a large contribution from the mountains of Ethiopia and the Blue Nile. However, opinions are divided as to whether climate is the only factor: the old Aswan Dam has also been cited as a possible explanation, albeit tentatively due to a lack of data, even though other studies had denied its role as a trap [FRI 91].

2.3.1.4. *Protection of the coast*

The 240 km of coastline are protected against the sea to varying degrees: 55% of the coastline is protected by dune bars and high sandy beaches, 30% is exposed with no defenses and 15% is protected by artificial structures (in particular the 20 km-long vertical wall that supports the coast road constructed in 1934 in Alexandria). On the basis of Muhammad Ali's instructions, boulder protections were constructed in the 1830s at the back of Aboukir Bay to dry out the lagoon; others were constructed in the 1980s to defend the Rosetta and Damietta promontories. The recurrent project of the Egyptian authorities for nuclear materials to use a 28 km stretch of the Borollos dunes, rich in zircon, monazite and ilmenite, is a concern for environmental specialists [HAN 13a].

2.3.1.5. *Salinization of waters and soils*

The water in the large channel has an acceptable level of salinity, since the inputs from the Nile, having drained its valley downstream of the Aswan Dam, are mainly composed of carbonates. Degradation of the water quality in the delta by pollution arises especially from the Cairo agglomeration. It also has wide-reaching and concentrated sources in the delta itself: agricultural drainage, industrial drainage (7.5 km^3/year) and sanitary drainage (6.5 km^3/year) produce used water that is released without treatment into the channels of the delta. Moreover, 25,000 tons of solid waste are poured each year into the 33,000 km of delta channels.

Resorting to the exploitation of underground water is important as much for quantitative reasons as for qualitative ones, but the limitations on this resource should activate the authorities. At the scale of the valley and the delta, underground water represents 20% of non-agricultural resources and 20% of the volume used in the irrigated perimeters [ABD 13]. The volume available on average per delta inhabitant for all economic and human uses is now only 660 m^3/year, meaning that underground water has become vital over the years. The salt concentration is not excessive in itself, with the exception of the coastal plain, where it is 3.2 g/L; it is effectively 0.16–0.5 g/L in the South Delta and 0.5–1.5 g/L in the Middle Delta. The increase in concentrations downstream is due to infiltration of irrigation water and to the intrusion of sea water, facilitated by excessive pumping. One of the major issues for the delta is the link that exists between, on the one hand, the very mediocre quality of its water for economic and human use and, on the other hand, excessive pumping from underground reservoirs.

Another issue for the future of the delta is the development of secondary salinization of soils, i.e. related to irrigation. According to the international classification of soils, those in the delta are clayey and salty or "altered by salt" in their northern, lowest, part and in the small sectors of the Middle Delta, where bad management practices can be incriminated, in particular the use of drainage waters for irrigation. The soils in the southern part of the Delta are not considered to be affected.

2.3.1.6. *Maintaining flow through to the sea*

What methods exist to counter the process of salinization? First, a solution would be to maintain a flow of fresh water to the Mediterranean. In 2000, an annual flow rate of 12.3 km^3 was still reaching the sea and the last lakes, produced almost in its entirety by the drainage of irrigated perimeters. Yet it has been demonstrated that a rate of 10 km^3/year flowing through to the sea is required to inhibit salinization of soils on the maritime edge of the delta. This value is used to determine the irrigation of the new irrigated surfaces of the coastal plain to the north of Sinai. The remaining 2.3 km^3 would be obtained by mixing recycled drainage water and fluvial water.

2.3.1.7. *The significance of rice cultivation*

Second, it is important to constantly improve rice cultivation in the part of the delta that is close to the coast. The delta is a rice area, a strategic culture, which is, however, prohibited by the state authorities in the valley of the Nile where it consumes too much water. In the delta, rice is cultivated in summer as a rotating crop on a 3-year cycle along with cotton and corn (with the second crop, in winter, being beans or clover, improving fertility by capturing nitrogen from the atmosphere and thus avoiding the use of fertilizers); depending on the years, it occupies an area of between 420,000 and 630,000 ha and offers the high yield of 80 qx/ha, with a very low risk of disease; the produce is destined for Egyptians and significant quantities are exported. Rice irrigation carried the great benefit of encouraging the process of washing salt away, but on the condition that the drainage can take place, i.e. the water table is at a depth of 2.5 m below the ground surface. The state authorities have put significant means towards increasing subsoiling practices, the use of ceramic underground drains and the use of gypsum to prevent alkalization of soils. At the scale of the producers, it has been shown that clover cultivation does not cause salt to rise, in contrast to cotton cultivation. However, rice cultivation has had negative effects on soil compaction and requires a large quantity of water, now set at 21,000 m^3/ha, to be fully effective with regard to salt accumulation by the effects of capillarity and evaporation. The authorities aim to cut the rice cultivation areas by half in order to prepare for water shortages and to reserve these areas for the coastal fringe where the effects are felt the most [KOT 00].

2.3.2. *The Rhône Delta: changes in the basin and the delta*

The Rhône Delta possesses the landscape of a Mediterranean delta. The 1,500 km^2 of the Camargue constitutes a "curious building", where, at the end of the 19th Century, very different landscapes were juxtaposed with other: that of extensive livestock farming of bulls and Camargue horses on the edges, that of industrial extraction of salt in the southeast marshes, a rich environment of marshes and lagoons behind the barrier beach, an area for wildlife, and finally agricultural terrain. From the 17th Century onwards, the deltaic plain of the Camargue was protected by dykes (locally named "chaussées" or embankment), irrigated by "roubines" or channels to the sea to wash out the salt, and drained and scattered with farms (locally "mas") in order to develop a winning agricultural economy. The requirement for capital meant that landowners were from outside the region and land was exploited by farmers, where the size of the property was proportional to the salinity level of the water; on this empty and insalubrious land, authorization was given for the use of technology to improve cereal cultivation and livestock prairies (sheep), then for irrigated wine-growing and rice culture. The Camargue was cut off from the Rhône after the 1840 and 1856 floods, but not from its irrigation waters. Agriculture came into conflict with salt producers, who refused fresh water and obtained a "dyke from the sea" to better secure the salt marshes against storms. The Vaccarès and the pools no longer evolved, due to a lack of sedimentary inputs, and were then transformed into a nature reserve by the state. Production activities and nature cause conflict between, on the one hand, the forces of progress, arriving from the north of France with vested interests in the exploitation of resources, and on the other hand, members of the elite classes from the south who seek to protect nature [PIC 88].

2.3.2.1. *Regression of the Rhône delta at the end of the 19th Century*

After going through a period of active growth during the Little Ice Age, the Rhône Delta began to regress. All of the values suggested until recently have suffered from great incertitude, due to a lack of knowledge of all the figures involved in the sedimentary budget and of the proportion of the bedload with respect to the suspended load; we must therefore content ourselves with rather approximate estimates. The inputs at the delta head (at Beaucaire) have been reduced from

17 Mt/year (perhaps less) to 5–7 Mt. The tonnage traveling in the channel as bedload has decreased significantly since the 19th Century. The channel had a braided morphology (with multiple channels featuring banks of bedload) in the middle of the 19th Century. The morphology of the channel of the Rhône Delta has progressively metamorphosed, in response to the reduction of solid transport, which itself is due to a combination of factors acting at the scale of the basin:

– spontaneous reforestation related to the depopulation of mountains, locally prior to the middle of the century, and the reforestation organized by the state after 1860;

– the construction of hydroelectric reservoirs on tributaries after 1920;

– entrapment in compartments known as "casiers Girardon" (1880–1940) along the Rhône and in the old side channels; finally, in the storage reservoirs constructed by the CNR (National Organization for the Rhône) after 1945;

– massive gravel extractions in the old braided beds; these have been significant since the 1930s, massive between the 1950s and the end of the 1980s, before almost ceasing at the beginning of the 1990s.

The factors reducing the sediment load of the river were activated at the scale of the watershed at the very moment when the Camargue was hermetically closed off. The reduction in the load imposed from upstream will not therefore affect the deltaic plain, but in fact the coast itself. The new metamorphosis of the Rhône is expressed by a narrower and more incised channel, as well as by aggradation of the banks. These changes explain why more than 90% of the channel materials reach the mouth. The inputs have therefore reduced in tonnage since the middle of the 19th Century, but are rarely deposited in the deltaic plain. Does the coast benefit from them, for all this? In fact, the greater part of sedimentary inputs reach the open sea without benefitting the coast; on the contrary, today, the coast has a slightly negative balance, because it loses materials that have been carried by long-shore drift. The materials in the bedload and the fine suspended sand, which are very useful for the defense of the coastline in the face of rising sea levels even when the quantities are reduced, are deposited at depths greater than 20 m, which is considered to be the limit below which they cannot be picked up by the waves and swell. This projection into the open sea is because the load is increasingly fine and dyking, which reaches all the way to the mouth, nearly no longer allows direct supply of the coast by the river.

A summary proposes the following values for the solid flow rate and the Rhodanian deposition environments in the middle of the 19th Century, i.e. before the crisis (Table 2.1).

Years	Upstream inputs (Mm^3/year)	Storage at the mouth (Mm^3/year)	Exit on coastal platform (Mm^3/year)
1870–1960	16–17	3.5	14
1960–2000	5–7	0.5	5–7

Table 2.1. *Summary of the sedimentary transfers of the lower Rhodanian system during two time periods. (Source: [ARN 03, MAI 07, PRO 14])*

The freedom left to the dynamic of the deltaic branches therefore belongs to the past. Sediment fluxes of the Greater Rhône when in flood are now almost entirely guided by the existing levees downstream from Beaucaire. Geographers in Aix-en-Provence in France have demonstrated that a significant part of the sediment reaches the continental platform and is lost to the coast, since their deposition depth on the deltaic front is too great and does not allow sand to be picked up by the waves and long-shore drift.

2.3.3. *The Ebro Delta: alone against the sea*

2.3.3.1. *Landscapes of the active delta*

> "At the mouth of the great river, in the sea itself, a strange amphibious landscape appears... The Ebro Delta... It progresses for around 10 meters per year at the Buda headland, thanks to the constant supply of enormous alluvial masses, which turn the sea yellow for up to 10 kilometers off the coast, as we can easily see from an airplane. These silty waters encourage a surprising amount of fauna, a true breeding ground which is rare in the Mediterranean, a sea relatively poor in biological terms" [DEF 57].

The delta was a hunting and gathering region, with extensive livestock farming of cattle that is reminiscent of the Camargue, within an unsettled and almost empty country, since "no delta civilizations had been established in the (western) Mediterranean".

The Ebro Delta covers an area of 330 km^2, which has prograded by 25 km into the sea. The delta has old gravel bars, two lobes and two sandy spits that isolate two bays to the north and the south. Its morphology shows that it was mixed, both river-dominated and wave-dominated, with the coastal front displaying significant redistribution of sandy input materials.

The delta remained marshy until the 1860s, when the practice of diverting water loaded with suspended sediments from the river was adopted. This was the era when the annual load of the Ebro was 28 Mt, from 1 to 1.5 Mt of bedload, with high concentrations between 1 and 10 g/L. The channels allowed uplift rates of the delta of 0.5 cm/year by diverting 5% of the flow rate and 0.5 Mt/year over an area of 200 km^2. The natural soils had aggraded to a total of 50 cm, to the point of being exempt from flooding, and increased their productivity. Creation of a fertile soil has thus allowed rice cultivation, setting up a landscape of green paddies in summer and scattered rice plants [IBA 97]. In the 1920s, the former displaced the alluvial materials carried in by the irrigation channels that had quickly became full (locally known as the "sequis"), leveled the paddies (locally known as the "tancals") and washed the soil in winter by using the drained water (locally known as the "desaigs"); the excess earth formed small islands used for vegetable cultivation and also as rice-drying areas [DEF 57]. An area of 77 km^2 of wet environments has been turned into a nature reserve, which was integrated into a nature park in 1983 (Figure 2.7).

Figure 2.7. *The Ebro Delta no longer progrades at its mouth at the Cape of Tortosa. Waves redistribute the sediments by creating the small El Fungar barrier beach to the north and the one shielded by the Punta de la Banya to the south. (Source: CNES 2008, Distribution Airbus DS). For a color version of this figure, see www.iste.co.uk/bravard/sedimentary2.zip*

2.3.3.2. *Impacts of developments in the basin in the Ebro Delta*

The Ebro basin covers an area of 85,500 km^2, with the river itself being 928 km long. The flow rate has reduced from 590 m^3/s at the beginning of the 20th Century to a current level of 425 m^3/s. The annual flow rate of the Ebro at the mouth is 12.6 km^3, but it would be higher if 6.3 km^3 were not taken up for irrigation and 3.35 km^3 for cooling of two nuclear power stations.

The basin has more than 190 storage reservoirs created by dams that regulate a volume of 7.700 hm^3, namely 60% of the average annual flow. Two-thirds of that capacity were commissioned between 1950 and 1975. Two large reservoirs were inaugurated on the enclosed course of the lower Ebro (Mequinenza in 1966 and Ribarroja in 1969). The degree of development upstream of the delta means that the Ebro is a good example of a "harnessed" watercourse.

The natural input of sediments had already been reduced from 28 to 5–6 Mt/year when the first dam was constructed on the lower Ebro, especially due to entrapment of sediments in Pyrenean reservoirs and forestation of the basin. The load entering the reservoirs of the lower Ebro has been reduced further, since it was only 0.5 or 1 Mt/year in the 1990s [BAT 03, BAT 11]. The materials deposited here are made up of about 90% silts and clays, which reflects the lithological characteristics of the watershed. What quantity is able to reach the delta today when these impacts are taken into account? By assuming an entrapment efficiency of 90% of the inputs, the dam complex on the lower Ebro allows a reduced supply to the delta of 0.2 million tons/year, perhaps even 0.12 Mt, where this figure also includes materials coming from erosion of the downstream section of the last dam. This is still only the suspended load. The bedload entering these reservoirs is effectively zero, because despite a significant transport capacity, the flood hydrology has been insufficient and dredging has been carried out in the river to meet navigation requirements as well as on the river and its tributaries that carry a stony load for aggregate supplies. It is indeed important to draw attention to the intensity of gravel extractions from Pyrenean tributaries such as the Segre. The solid flow rate that passes through the series of dams is now equivalent to 1% of the solid flow rate estimated for the beginning of the century. The sediment deficit is therefore considerable and similar to that of the Nile, albeit in a very different context. The Ebro, a less abundant river than the Rhône, but that has perhaps transported more sediments, has lost half of its water and is a good example of near elimination of the solid flow rate of a Mediterranean river originating from nearby mountains, in the space of only a few decades.

2.3.3.3. *A delta subject to wave action*

The morphology of the current delta feels the effects of the significant sediment deficit of the Ebro; the waves, which have become the only agent of their evolution, soften the trace of the sandy coastline to the expense of the former active lobe and cause faster recession of the coastline. Since the 2000s, storms have also created long-lasting breaches through the thin and low-lying sand barrier beach; overflows of sea water through these breaches modify the salinity and ecology of the lagoons [JIM 11]. The margins for maneuver that are still possible remain to be seen; in recent years, the practice of experimental flushing has been introduced, with attempts to remobilize some of the deposited sediments (see Chapter 5).

2.3.4. *The delta of the Po plain: historical dispersion of weak points*

2.3.4.1. *Accelerated regional subsidence*

One of the first areas on Earth where accelerated subsidence has been studied – and, we should note, halted – is the lower plain of the Po, both fluvial and deltaic in nature. High stakes are at play, even though the amplitude of the sinking is limited in comparison to other regions of the world, insofar as historical cities fear for their heritage, as coastal ecosystems of great value are threatened by the rise in sea levels and finally as the risk of flooding has increased significantly. The fact is that the majority of the area covered by the Po Delta is below sea level, sometimes to the tune of 6–8 m. Natural subsidence proceeds at a relatively slow pace. Estimated to be 1.3 mm/year since the rise in sea levels, it is attributed to the subduction of the Adriatic plate under the Apennine Mountains range (1 mm/year on a very long-term basis); subsidence is doubled by the effect of the rise in level of the Adriatic Sea [CAR 05].

Here, as much as elsewhere, "accelerated" subsidence has taken place on an entirely different scale. Acceleration by human actions has been caused by land drainage to aid agricultural development, by intense water pumping carried out to accompany economic development of the Po and the Venetian plains, and finally by extraction of methane since the end of the 1940s. Peak subsidence was reached between 1951 and 1957, with considerable values of 18–25 cm/year between 1958 and 1962, before dipping below 4 cm in the early 1970s. Thanks to a halt in gas extraction, the movement slowed down significantly, to the point of reaching today's level of approximately 5 mm/year, but the rise in sea level has required specific measures to be put in place on the coast [FAB 14].

2.3.4.2. *Subsidence of the Venetian Lagoon and defense of the coastline*

Constructed between the mouths of the Piave and the Adige, a sandy barrier beach (known locally as the Lido*) of about 40 km in length protects the Venetian Lagoon and the old historical city. The Lido has always been at the mercy of the storms that affect the Adriatic Sea; today, it is receding away from the sea, and its defense structures (the *murazzi*) are sometimes topped by the waves, such as in 1966.

Subsidence of the Venetian Lagoon is due to pumping of water from artesian wells in the Mestre industrial area and on the Lido to meet the requirements of the tourism activity that has developed there. It reached a culmination point at the end of the 1960s, with values close to 10 mm/year in Venice, then it slowed down dramatically, before briefly rising again at the beginning of the 1970s. Since then, only the natural settlement of the soil is involved (0.4 mm/year) along with localized anthropological effects that give rise to a rate of 1.6 mm/year. The current difficulties lie elsewhere, in particular in the threats that hang over the ability of the Lido to remain intact, posed by the halt to sedimentary inputs from the Piave and by the deficit recorded in long-shore drift; this can be seen in front of the *murazzi*, where the slope of the submerged coast has steepened under the effect of marine erosion, in particular during successive storms [GAU 17]. Most obviously, it is subject to a direct threat from tides, corroborated by the rise in sea levels. With respect to the zero level of the height of tides as measured at the La Salute promontory, the *Acqua alta* begins at the + 80 cm level, for which St Mark's Square would be inundated; the exceptional level of + 140 cm has been exceeded three times in the last 50 years (at the level + 200 cm, 90% of the city's land area would be underwater). To mitigate these threats, the three wide breaches opened in the Lido are in the process of being equipped by the dykes, for which the project was initiated in 1970, following the record *Acqua alta* of 1966. The giant MOSE[5] project has already cost 6 billion euros, and the cost will increase by the time it is completed if it is completed one day.

2.3.4.3. *Ravenna and Bologna*

Another affected area is the Ravenna region, 60 km to the south of the mouth of the Po River, where subsidence, which locally exceeds a cumulated value of 1.5 m, is due to pumping of potable water from the upper aquifer since the end of the 1940s; this was aggravated from 1952 to 1961 by the extraction of methane from the Ravenna Terra gas field, located below the plain in the folded Pliocene layers. Modeling has shown that the contribution of pumping of potable water exceeded the extraction of gas by a few decimeters. At the end of the 1970s, undesirable effects affected monuments, the industrial zone and the marshes; the watercourses that cross the center of town now pose a risk of flooding.

Subsidence that locally exceeds 1 m has also affected the region of Bologna since the end of the 1940s, due to the extraction of water from the deposition fans of the torrential rivers that drain the Apennines; since the fans were connected to each other, the lowering of the water table affected the intermediate terrain formed from

5 A total of 78 floating dykes are planned, over a total distance of 1,600 m; they can be lifted up by tides of 110 cm and above (12% of the surface area of Venise is then flooded). MOSE is the abbreviation of *Modulo Sperimentale Ellettromeccanico* (Experimental Electromechanical Module).

fine and highly compressible materials. In the center of town, consecutive measurements with damage recorded on old public buildings have demonstrated that subsidence reached 10–80 mm/year. The process has been controlled since the end of the 1980s due to a reduction in pumping [BAL 91, CAP 91].

2.3.5. *The Danube Delta: still room for hope*

The Danube Delta, which is the second largest delta in Europe after the Volga Delta, covers an area of 3,500 km^2, which has developed in the Black Sea at the mouth of a basin with an area of 817,000 km^2. It has a very great environmental and cultural value, and has the title of Biosphere Reserve; its archeological riches include very large numbers of sites dated almost continuously from the Neolithic era to the Middle Ages.

2.3.5.1. *The components of the delta*

The interior delta formed around 9,000 years BP in the lower river valley in the form of a vast *embayment**, and then the lobes of the external delta developed from 7,200 years BP onwards. Today, the delta has three main arms, which each supply their active lobe (themselves described as deltas): these are, from north to south, Chilia III, Sulina and Saint-Georges II, which are encased in a vast collection of lakes and lagoons, marshes and sand bars (Figure 2.8). The abundance of abandoned channels is a characteristic of this delta, where diffluences have been frequent. The well-fed lobes progress, whereas the lobes placed at the mouth of weakened channels (Sulina) are subject to wave dominance and long-shore drift from the northeast; they recede, which benefits several successive generations of coastal bars. The oldest, visible in the deltaic plain, are locally shaped into dunes by the wind [FIL 14, GIO 05, PAN 12]. The Chilia lobe was reactivated by a partial avulsion until the medieval and modern development of the Chilia III lobe, which reached its maximum activity around 1700. Progression was rapid during the Little Ice Age due to the climate and because clearing affected the watershed: this is due to late deforestation of the Lower Danube basin, erosive rain and large floods. Was deforestation caused by the demographic growth that corn crops have permitted? A different hypothesis exists: a very high demand for ovines, started by the Ottoman occupation for the benefit of Istanbul, in hills and mountains that had until then been dedicated to other activities [FIL 14].

2.3.5.2. *Alterations of delta equilibrium*

The Danube Delta underwent successive impacts: channeling of the Sulina arm to reduce navigation time (1868–1902), and then large improvement projects after 1945, in the mindset of economic development in the communist era. Most importantly, this included the connection channels for activities at an industrial scale: fishing, fish farming and harvesting of reeds, and then construction of agricultural polders (950 km^2).

It is important to put the very costly Danube–Black Sea canal (96 km) in a class of its own. This was dug by political prisoners between 1949 and 1984 through the Dobruja, between Cernavoda (a port on the Danube) and the sea port of Constantza. Tourism has taken over after the abandonment of the socialist-type economy in favor of activities involving the discovery of the ecology and the delta landscapes. One of the ecology issues is to recreate the conditions that allow sturgeon to swim upstream beyond the Portes de Fer dams to reach Szigetköz (Hungary), the natural zone short-circuited by the Gabcikovo dam (Slovakia). However, the European policy on transports, which, in the same way as those of neighboring states, aims to increase the navigational channel dimensions of the watercourse, constitutes a serious obstacle; the solution lies in a compromise between the request in terms of transport and protection of the alluvial plains that are at risk of lowering the groundwater table.

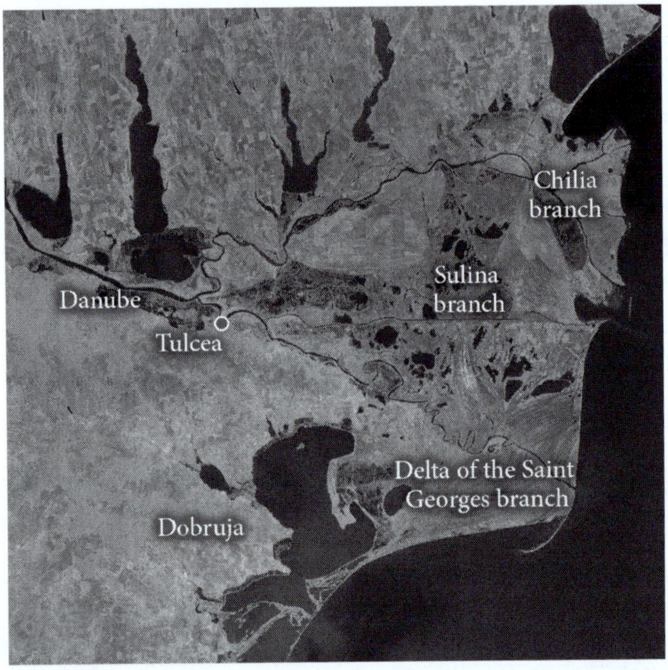

Figure 2.8. *The Danube Delta spreads between the Bugeac hills to the north and the Dobruja to the south. The external delta deploys from north to south: the narrow Chilia arm, with multiple mouths, that constructs the lobe of the same name; the mouth of the Sulina arm in the west-east direction; finally, to the southwest, a vast collection of lagoons is protected by the barrier beach formed by the long-shore drift, moving in a southwesterly direction, that carries the alluvial materials brought in by the Saint-Georges arm along its winding path. Old coastal formations can be made out in the internal delta. For a color version of this figure, see www.iste.co.uk/bravard/sedimentary2.zip*

2.3.5.3. *Justifications for hope?*

Concerning maintenance of stability of the delta, two opposing trends have come to light. On the one hand, incision of the network of channels has encouraged dispersion of sediments in suspension in the delta (but modestly so: 2.5 Mt/year), whereas the solid flow rate has fallen due to the construction of dams at the scale of the basin. From upstream to downstream, these are the dams on the Danube between Poikam and Jochenstein (2,400 and 2,200 km), then the Gabcikovo dam (1,842 km), and finally the Portes de Fer I dam (942 km, in 1970) and the Portes de Fer II dam (846 km, in 1983). Since the load arriving in the delta decreased from 75 to 25 Mt/year in the second half of the 20th Century, the sandy component is no longer greater than 4–6 million tons.

A positive point is that internal construction of the delta has continued at the same pace of 2.5 Mt/year, despite the significant reduction in solid transport from upstream; this construction removes 10% of the fluvial solid flow rate at the exit, which is quite rare on a global scale. The fact that sedimentation on the deltaic plain has reached an average of 3 mm/year makes researchers "prudently optimistic" about the behavior of the delta regarding the rise in sea levels. In fact, the effectiveness of the channels is reducing and the prediction of lower flow rates in the future will go the same way. However, the coast is receding because the deficit is significant at the mouths; the sediments carried in by the main branch of Chilia are redistributed along the coast in a southerly direction and threaten the mouth of the Sulina branch; in addition, the dredged sand is dumped off the coast without having any benefit for the coast. The coastline has been receding for centuries to the south of the Sulina branch, because supply to it is insufficient.

A series of human actions have certainly made the deltaic plain more artificial, but they have somewhat reproduced (or imitated) a network of natural distributaries; by facilitating the entrance and dispersion of the flood waters, they in fact encourage the construction of the deltaic plain [GIO 13, PAN 98]. The delta has benefitted from the status of biosphere reserve since the fall of the communist regime (1990). Restoration programs have been put in place for 15,000 ha since 1994 and regularly renewed, which should allow the evolutions in progress to be accompanied, in particular the rise in sea levels.

2.4. Conclusion

In the long and middle terms of the Holocene, the Nile provides an example of a delta that is subject to natural fluctuations in hydrosedimentary input from its watershed, whereas the Huang-He is subject especially to fluctuations in sedimentary input controlled by development on the highly sensitive Loess plateau; the two rivers are strongly contrasted in terms of determinism of their evolution.

At a modest scale, but with a complexity that reflects the multiple aspects of the watershed, the Rhône Delta has recorded the intersecting influences of climate and human action, as much in the dynamics of its channels as in that on the coastline.

The delta of the Rhine and the Meuse is, along with the Po, the most occupied delta in Europe. Dutch society has, for at least eight centuries (Volume 1, Chapter 1), fought against fluvial flooding and the sea. The contemporary era has seen almost total control of flooding and a sediment deficit related to the isolation of low-lying areas that are subject to subsidence and the rise in level of the North Sea. This delta is also the site of remarkable experiments at the crossroads of total protection policies with the Delta Plan and the recent desire to maintain environmental quality. It has become the epitome of a controlled delta.

Moving on to the present day, we presented last of all the recent trend of several deltas in the old Mediterranean world that present an imbalanced situation. All share the common factor of experiencing significant hydrosedimentary fluctuations of a fluvial nature, in particular sediments have very markedly dried up in the case of the Nile, the Rhône and especially the Ebro, drying-up that no longer allows the deltas to fight with equal weapons against subsidence and the rise in levels of the Mediterranean. Only the Danube, which suffers less impact from dams and has distribution channels, retains the potential for compensation in the face of a sediment deficit.

In Chapter 3, we will consider the situation of tropical deltas in the face of change.

3

Tropical Deltas in Crisis, Between Open and Closed Formations

This chapter is dedicated to deltas in developing countries that are highly exposed to crises and change due to their open angle to the ocean, their low level of structural protection and their often dense and highly vulnerable population.

We will successively look at the cases of the Ganges–Brahmaputra, the Mekong and the Niger, all deltas that are alive but under threat to various degrees; then, we will examine the case of the Indus Delta, subject to catastrophic drying-up caused by humans that have made it resemble Colorado but in a very different context. Lastly, the future of the Irrawaddy (Ayeyarwady), perhaps the most natural of the large deltas, remains on hold, but is in the process of undergoing profound transformations.

3.1. A delta that is both open and alive: the Ganges and Brahmaputra Delta

The watersheds that control the delta of the Ganges, the Brahmaputra and the Meghna have remained relatively natural up until recently. The fluvial flow was controlled by a reduced number of hydraulic structures, and the vast delta at their mouth had, and still has, very little protection from the Indian Ocean.

The region is affected by two monsoons separated by a short dry season:

– the summer monsoon (from the southwest) waters the Indian subcontinent from June to September;

– the winter monsoon brings water to the region from December to March (arrives from the northeast);

– lastly, there is a difficult period of drought and shortage from April to May.

The two monsoons cause high water levels, the summer monsoon may be accompanied by cyclones that form in the Indian Ocean and lastly, the short period of dry weather is the subject of international disagreement in countries with few water reserves. A deciding factor, the Himalayan mountain range provides a very abundant sediment load, but the development of the upper Brahmaputra has begun and has ceaselessly been a problem.

3.1.1. *Rivers and a delta*

The Ganges Delta (so-called under the old English denomination) is now known as the Brahmaputra–Ganges–Meghna delta. The largest on Earth, it covers an area of 105,000 km^2 if the active delta to the east and the old delta to the west are taken into account. The delta in the wide sense of the term encompasses waters and sediments from the three large rivers, whose basins accumulate a surface area of nearly 1.7 million km^2. The hydrographic network is all the more complex since the course of the rivers has varied several times throughout history (Table 3.1):

– the Ganges, before entering Bangladesh, has an annual liquid flow rate of 490 km^3/year. A little before its confluence with the Brahmaputra, it takes on the name "Padma". Over its course, it loses water to numerous distributaries* on the right bank;

– the Brahmaputra (which bears the name "Jamuna" in Bangladesh) flows with an annual flow rate of 630 km^3/year;

– after the arrival of the high Meghna at the Padma, the river again changes name and becomes the "lower Meghna", which finally empties out into the Bay of Bengal (Table 3.2).

Throughout its Indian route, the Ganges loses some of its waters to the benefit of a distributary, the Bhagirathi–Hooghly, which runs through Kolkata (known as Calcutta until 2001) and forms the border between India and Bangladesh. This watercourse was the main channel of the Ganges until an avulsion that occurred in 1789; therefore, this event was at the origin of the current course of the Ganges that passed through Bangladesh.

Rivers	Basin surface area (km^2)	Liquid flow rate (km^3/year)	Solid flow rate (Mt/year)
Ganges	907,000	490	520
Brahmaputra	640,000	630	540
Meghna	80,000	150	

Table 3.1. *Liquid and solid flow rates in the Ganges–Brahmaputra–Meghna system [MIL 13]*

The Brahmaputra part was very significantly strengthened with respect to that of the Ganges in the years 1980–1995 [ISL 99]. Table 3.2 distinguishes the sediments that are deposited in the delta (plain and channels) and the component that truly reaches the Bay of Bengal.

Total load	Deposited part (Mt/year)		Part reaching the ocean (Mt/year)
1,037 Mt/year	512 (49%)		521 (51%)
	Alluvial plain	289 28%	
	Channels	223 21%	

Table 3.2. *Load and deposition sites for sediments in the Ganges–Brahmaputra system [ISL 99]*

Moreover, the Hooghly is believed to transport 328 Mt/year, such that the total suspended load arriving in the Gulf of Bengal is approximately 850 Mt/year.

3.1.2. *The Ganges–Brahmaputra–Meghna plain, the most populated and the poorest on Earth*

Bangladesh has more than 180 million inhabitants over an area of nearly 90,000 km^2, which gives the country a human density of 1,230 inhabitants/km^2. The delta covers 88% of the area of Bangladesh and houses 110 million inhabitants (including its Indian part, the delta is populated by 170 million people, primarily farmers). The lands and the jungles in the delta were colonized in the 19th Century by "pioneer-settlers" who were concessionaries and subtenants of lower status. The soils in the "moribund" delta to the west were quickly exhausted and turned out to be malarial (Figure 3.1). The demographic explosion in the 20th Century, in addition to property and social inequalities, found a release in the colonization of "chars", which are low-lying and mobile lands, young and fertile, that belong to the active delta; these are islands or lands next to the banks, which are physically unstable and on which great demands are placed by colonization, despite their susceptibility to flooding. The chars are very dangerous environments and have a very uncertain lease status due to their intrinsic mobility, to their floodability by river branches or by cyclones, and to the instability of the legislation that applies to them; conflicts that result from this sometimes involve bloodshed. These lands – pastures and fishing grounds at first – were developed for the cultivation of rice paddies and indigo at the end of the 19th Century. Nowadays, more than five million inhabitants live a desperate existence in the active delta, because it is the land of some of the poorest people. When the land was in the process of consolidating, pioneering rice cultivators were chased away by leaseholders who had obtained their leases under the cronyistic system and via new speculators, namely shrimp producers, in a region where a kind of anarchy reigns [CHA 06, MAL 12].

Due to a lack of vegetable proteins in sufficient quantities, the population is increasingly turning to the consumption of fish, but the quantity of proteins available is constantly reduced due to the reduction of biomass and of the diversity of natural species. Landing of fish caught in the natural environment was reduced by a third between 1990 and 2010 due to pollution and unsuitable practices. This loss, which is not confined to Bangladesh, is now more than compensated for in volume by aquaculture products and activity that is developing at a rate of 8% per year (the country is the sixth producer worldwide). These products, comprising very large species, are a valuable source of protein; however, they contain fewer micronutrients (vitamins and oligoelements) than small indigenous species. The nutritional balance is in fact negative for one of the most underfed populations in the world. Malnutrition is exhibited by the fact that a third of children below five years of age suffer from rickets and millions of people suffer from a deficiency of micronutrients. One solution, still not widespread enough, would be to develop a polycrop associating vegetable proteins and some indigenous species that contribute vitamin A to the diet [BOG 17].

What is the explanation for this kind of situation? This is due to the fact that the Ganges basin (and this statement can be considered as identical with respect to the Brahmaputra) is one of the most underused river systems in the world from a hydraulic point of view, even though its water resources have been overexploited, as much for agriculture as for industry. This system combines an excess of water and extremely severe deficits in a dry period.

3.1.2.1. *Catastrophic flooding*

Floods are the most obvious manifestation of the amplitude of waters rising caused by the summer monsoon that fall on the Indian subcontinent, in a catastrophic manner in some years. Bangladesh and the Indian part of the delta are floodable lands that affect one of the poorest regions of the planet. 20–30% of the surface area of Bangladesh, over which most of the delta extends, is seasonally floodable, by rivers and by the sea: these are fluvial floods from the monsoon season (*kharif*) and pluvial floods (torrential and from the Eastern range) from June to October, and lastly flooding linked to tides and storm surges from the sea during storm waves. In the event of a disaster, 80% of the country can be flooded (Figure 3.1).

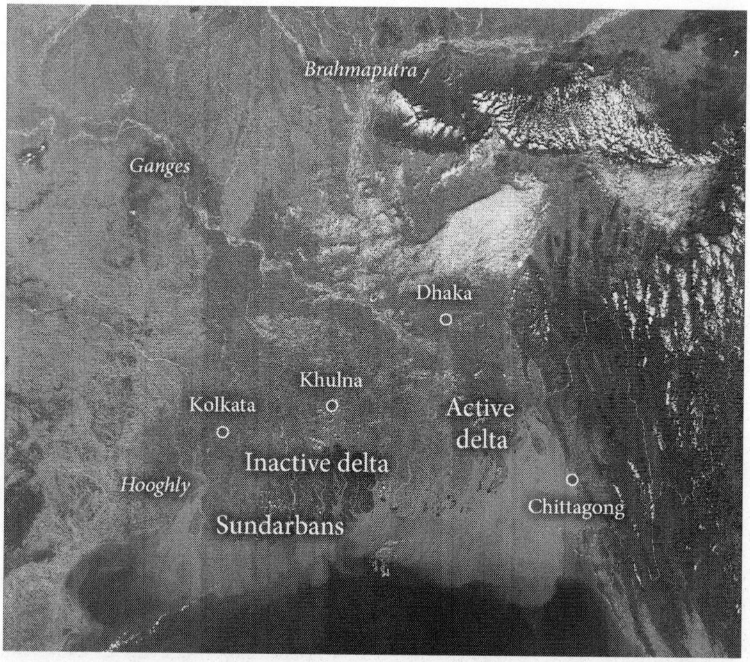

Figure 3.1. *The delta of the Ganges and the Brahmaputra. Hydrographic network: 1. Ganges; 2. Jamuna; 3. Padma; 4. Meghna; 5. Hooghly. (Source: Jacques Descloitres, MODIS Land Team). For a color version of this figure, see www.iste.co.uk/bravard/sedimentary2.zip*

The Bay of Bengal is subject to cyclones, which raise the sea level and invade low-lying areas of land. Since 1737, 23 disastrous episodes have each killed more than 10,000 people, in particular in 1737 (300,000), 1864 and 1876 (100,000), 1897 (175,000), 1970 (300,000) and 1991 (140,000 deaths and 10 million homeless). Climate change may be responsible for the increase in the frequency of flooding, even though it is slightly too early to confirm this. The long duration of flooding (two months on average) partly explains the significance of the damage, in the range of $2–8 billion for the 2004 event (more than 4% of the GDP).

The Ganges and Brahmaputra floods take place during the summer monsoon and are serious when they occur simultaneously. We can summarize the course of the last 50 years as follows: every two years, 20% of the country is submerged; Bangladesh has been through five fluvial flooding events, responsible for submersion of a third of the country, whereas the catastrophic floods of 1988

(Chapter 5) and 1998 (the *bonna*) flooded 60% of the country. It is true that Bangladesh has constructed thousands of kilometers of levees and dug drainage channels, but when the earth levees give way, they make the flooding worse. The importance of non-structural measures, or at least the attempt to generalize them, arises from an analysis: dykes are not deemed to be effective because they can break and, if this occurs, they close off the space, slow down drainage and reduce sedimentation at the surface of the deltaic plain; even rice cultivation has this negative effect [AHM 06]. In addition, the country does not benefit from good coordination with India, which only provides its information one day in advance, or not much more than that.

The geopolitical dimension of floods in this region of the Earth can be expressed during exceptional events. The World Bank intervened from 1963 onwards in the context of projects that reconcile protection and development of irrigation. The scale of the issue has changed in favor of large-scale protection of the alluvial plains of the Ganges, the Jamuna and the Meghna after the 100-year flood in the month of August 1988 (30 million homeless, between 1,500 and 2,000 deaths). A proposal was presented by the French president François Mitterrand at the United Nations General Assembly during the autumn of 1988:

> "Development takes place through large projects of global interest that are capable of mobilizing energies to be used in such and such a region that has been affected by nature or the madness of mankind. The example of the stabilization of the rivers that flood Bangladesh, at the origin of an impressive catastrophe, would be deserving material for a first project of this kind".

F. Mitterrand, encouraged to act by his wife who had been present during the summer 1998 floods, made a request to his advisor Jacques Attali to be in a position to make a great announcement at the Arche Summit in Paris, a summit of large industrialized countries that coincided with the 200-year anniversary of the French Revolution. Given this mission by France, Europe and the World Bank, J. Attali set up five design offices, notably the BCEOM (Central Bureau of Overseas Installations) and the CNR (Compagnie nationale du Rhône), under the umbrella of the FEC (French Engineering Consortium). The pre-feasibility study, financed by French public funds and completed in just five months, envisaged the construction of dykes between 3,300 and 4,000 km depending on the variants proposed, without any environmental impact study nor agreement with the countries in the basin, all of this in the specific context of corruption and regional violence. The Sommet de l'Arche brought together Germany, Denmark and the Netherlands under the coordination of the World Bank. The FAP (Flood Action Plan) phase of the study was to last for five years and take inspiration from the method known as "compartmentalization" that had already been tested on site by the Netherlands, a less ambitious method than that used in the plan by the French, and implemented from 1995 onwards. A review of criticisms of the FAP was presented by Dutch expertise, in parallel with strong recommendations in favor of an integrated view of development; displacement of the population could rise to the excessive figure of five million people according to one source; fertilization of land by small- and medium-sized floods would cease behind the dykes and aquaculture farming production would collapse, according to those with in-depth knowledge of the country [BRA 10, DUM 06, NDC 93, NIC 92]. A study carried out by the CNR planned the construction of concrete dyke elements to protect the banks, in places up to 100 m in height, in fluvial concavities, which shift laterally at high speeds. Not much was in fact carried out under the FAP from 1995 onwards, but the Dutch method was implemented (see section 3.1.2.5). This was the subject of criticism in itself, so much so that the newly elected government of Bangladesh modified the country's policy in favor of a National Water Management Plan based on projects of a more limited size and in principle based on more coordination at the local level [BRA 10].

Lastly, let us consider the example of the serious flooding caused in 2007 by the cyclone Sidr on the deltaic island of Mazer Char (Pirojpur district, 330 km southwest of Dhaka). This forested island, rich in fish, houses a few hundred inhabitants who live in conditions of extreme vulnerability. Poverty is so abject, and the inhabitants depend so much on their few possessions – in particular, their house and their livestock – that the evacuation advocated by the authorities is often refused in favor of an impossible fight *in situ*. This attitude, which is juxtaposed by that of populations who are sufficiently well-off to be able to give preference to their lives rather than protecting at any price the few possessions that will be essential to their survival after the crisis, should be taken into account in the risk management policies of these regions [AYE 17].

We have explored the issue of flooding, but this point must not mask the fact that the serious nature of hydrological events is not solely a meteorological phenomenon; it is at the heart of a complex web of factors that interact in an exceptionally serious manner.

3.1.2.2. *A subsiding delta, subject to the rise in sea levels*

From a geological point of view and at a very small scale, the Indian shield is overlapped to the north by the Himalayan mountain range and to the east by the Indo-Burmese arc in the contact of a double subduction zone. The Himalayan mountain range has been formed since the Eocene Epoch (35 million years), and the sedimentary fluxes of rivers have created an accumulation with a thickness of more than 16 km on the oceanic crust. The eroded materials accumulate in enormous quantities on the plate of the Indian subcontinent and advance towards the subduction zone which has been subducting to the east of Sylhet under the Indo-Burmese tectonic arc since the Pliocene Epoch. Based on a very large quantity of data, subsidence calculations yield a rate of 7–12 mm/year in the Sylhet basin, whereas the coastal band subducts at a rate of 3–7 mm/year. The rate of subsidence has recently increased and expresses itself differently in spatial terms, due to the multiplication of its causes, in particular of an anthropological nature (such as the effects of irrigation in agricultural areas and pumping from below towns such as Kolkata and Dhaka) [BRO 15]. To add to the complexity, measurements made with GPS systems have demonstrated that the weight of water stored in the plain during the monsoon period (more than 100 km^3) causes the level of the ground surface to vary from 20 to 50 mm [STE 13]. In principle, alluvial deposition compensates for subsidence. An archaeological approach produced interesting results in the Sundarbans mangrove. Salt ovens constructed during the dry season above the level of high tides, at least 300 years ago, have their base at 1.55 m below the current tide levels and are flooded during a catastrophic event. The rate of the rise in waters in the Sundarbans is thus estimated to be 5.2 mm/year on average, combining subsidence and the rise in sea levels [HANE 13].

Lastly, Bangladesh is very sensitive to salinization of surface waters, land and underground waters. The current dynamic strongly affects the spatial distribution of species and favors the establishment of vegetation with a preference for freshwater and briny habitats. One very serious consequence is the reduction in yields of rice cultivation. The rise in temperatures is added to this risk, which is likely to cause an estimated reduction of 8% in the production of rice and 32% in the production of wheat in the middle of the 21st Century [UNI 11].

3.1.2.3. *Sedimentary compensation of fluvial origin*

The solid flux that reaches the delta during floods caused by the summer monsoon is approximately one billion tons/year, in other words 10% of the sedimentary inputs from continental areas of the Earth into the oceans. A historical study shows that since 1792, the subaerial delta has actively prograded in the zone around the mouths, with an average gain on the Bay of Bengal of 7 km^2 per year [ALL 98]. In the eastern part of the emerged delta, channels, split bars and extended islands progress across the muddy subaqueous delta; on the other hand, the western part of the delta, deprived of sediments by the migration of river mouths, recedes by 1.9 km^2 per year. The delta is river-dominated and tide-dominated (with an amplitude of 4–5 m). In the area of the deltaic plain that is subject to an active fluvial influence, the average rate of accumulation is 23 mm/year (deposition of 300 Mt/year, with the remaining 700 Mt/year reaching the subaqueous delta). In the actively tide-influenced Sundarbans region, the accumulation rate is 11 mm/year, with an input of marine origin that largely compensates for low subsidence and confirms the ability of the mangroves to respond to the rise in sea levels.

The rate of rising of the sea level is greater than the global average (1.7 ± 0.2 mm/ year), which highlights the importance of subsidence.

3.1.2.4. *Traditional irrigation*

Born in India in 1852 on the edge of an irrigation canal, Sir William Willcocks paid homage to the ancient irrigation practices of the Bengal region [WIL 30]. Set up 3,000 years ago, after the Indo-European conquest, this system rivalled those of Egypt and the Tigris-Euphrates, which were approximately 1,000 years older; it was destroyed by the wars that led to the collapse of the Mongol Empire in the 18th Century. Thanks to the practice of *overflow irrigation*, the floodwaters of the Ganges and the Damodar added mud and fertility to summer monsoon waters. The principle, doubtlessly influenced by the Persians and the Arabs, centered on dams and reservoirs, by means of digging out shallow, well-spaced and inter-parallel channels, which transported mud during the flood peak; these channels that punctuated the upper level arteries systematically supplied the rice paddies and a diversified agriculture through breaches that were temporarily opened in their banks. The objective was to regulate rich deposits and enable malaria to be avoided, thanks to young fish from the river consuming mosquito larvae. This system, which covers 3 Mha, was not understood at the beginning of the British Raj (from 1858 onwards), but it was progressively restored in western Bengal, in particular under the influence of W. Willcocks, according to the principle of "living with floods".

3.1.2.5. *Responsibility borne by "modern" agricultural developments*

Breaking from the partially restored traditions, a modern system has been installed. The change took place in the 1950s, when a new governing body was established on the triple basis of Flood Control, Drainage and Irrigation or FCD/I. The question of land development to increase agricultural production was reformulated with regard to flood management.

In the 1960s and 1970s, the World Bank financed the Coastal Embankment Project, with the underlying principle of constructing earth levees with the objective of creating agricultural polders encompassing large areas in the former islands located behind the Sundarbans; the land was deforested, compacted, irrigated and drained, but deprived of sedimentary input, while alluvial deposition continued outside of developed areas; a relative levelling of 1.40 m was reached within approximately 40 years, to the detriment of these first polders [AUE 15]. The consequences of alluvial deposition and the relative lowering of the surface conquered by the polders are not the only ones. The development method launched in the 1950s has also increased dependency on added chemicals (subsidized to a small degree), because the soils become exhausted when they do not receive the fertilizing floodwaters; this leads to a concentration of property ownership among the richest, an increase in the number of landless farmers and a decrease in salaries, and lastly, an increase in the severity of famines due to reduced food availability.

We have seen that the 1988 and 1989 floods were motivating factors for implementation of the FAP (Flood Action Plan) in Bangladesh; a retrospective study on this subject has just been carried out on the CPP (Compartmentalization Pilot Project) for the Tangail district, to the east of Jamuna [RAM 18]. Between 1991 and 2000, a pilot project for 130 km^2 redeveloped an area that has been protected from floods arriving from upstream by a permanent dyke in the shape of a horseshoe, open in the downstream direction, by dams and waterways that allow water arriving from river branches and draining from land to pass through. The inflow of subsidies from the Netherlands and Germany to an area that presented so many similarities with the Dutch countryside has led to the neglect of critical points of view, in particular regarding the low level of implication in the projects by farmers, the low level of use of hydraulic installations, the reduction in transportation via unmaintained waterways and the fall in fish-farming production due to juveniles originating from the river no longer being dispersed, and even conflicts over water; the only obvious benefit appears to have been the shelter provided by the roads during floods. Implementation of FAP methods was stopped in 2000 by the government of Bangladesh, in favor of a new policy giving priority to opening up the land, as advocated in the *tidal river management* concept.

The same social criticisms are applied to the Blue Gold initiative, managed jointly by the governments of the Netherlands and two administrations in Bangladesh (the Water Development Board and the Department of Agricultural Extension). Over the period 2013–2019, 21 polders in coastal areas are to be revised and secured, for an area of 1,150 km^2, with the objective of training local managers at different territorial levels and ceasing all technical and financial external intervention by 2020, to give responsibility back to the communities. Rammelt *et al.* [RAM 18] are highly critical insofar as the local authorities have demonstrated their inability to exchange with poor farmers, who in turn feel that they are not listened to and that they have been wronged compared to producers who are capable of producing a surplus; they consider it highly doubtful that increased wealth will benefit everyone, since the poorest are employed in manual work. The factor attracting condemnation is more precisely the fact that Dutch cooperation has been placed under the umbrella of the Minister of Commerce for the Netherlands, in an approach that resembles depoliticized business, in which the main actors are rich farmers who are supposed to bring the poorer classes along with them; this approach resembles "surface flow", with growth taken hold of by the richest and then benefitting the poorest. Blue Gold "requires strong involvement of the private sector and will make good use of the expertise and know-how of Dutch enterprises and Dutch knowledge institutes, a first step in the transition from aid to trade”[1].

In summary, the same mistakes have been repeated for centuries at the cost of the poorest classes, non-commercial activities and the inclusion of minority groups. These protected areas contrast with the opening out of the Sundarbans (10,000 km^2), where sedimentation freely takes place in a regenerated mangrove and maintains an elevation higher than the average high water level of the sea (the polders have an elevation of just +1.5 m). The delta is therefore “relatively robust” in the face of subsidence and the rise in sea levels, despite the very obvious impact of human activities. Sedimentation, which takes place in the Sundarbans and in the active delta, allows surface losses due to marine erosion and the creation of areas of open water to be avoided. To restore the badly designed agricultural zones, coordinated creation of breaches is recommended in order for agricultural polders to also benefit from sedimentary inputs from tides and rivers and so that the surface of the deltaic plain is raised in a more homogeneous manner [BRO 15]. In this respect, the delta differs from the Mississippi and the Nile deltas.

3.1.2.6. *“Sharing” the waters of the Ganges*

Access to freshwater, for both irrigation and human consumption, is a huge challenge for India and Bangladesh, the western part of which is dependent on the waters of the Ganges. Issues surrounding quantities of water in the dry season, its quality and pollution are chronic.

The Bhagirathi–Hooghly supplies and drains into Kolkata, whose large port is located more than 150 km from the sea. Since the diffluence of the Ganges, the low flow rate of this watercourse, which has taken on the status of a simple distributary of the Ganges, has clearly caused sedimentation downstream: high waters bring in 10 Mt/year of silt, whereas the fluvial power can only evacuate 6 Mt/year, an extremely difficult situation, in particular during the dry season. In order to save Kolkata by increasing the flow originating from the river, from 1961 onwards, India studied the solution of damming the Ganges at Farakka (on Indian territory) in order to divert some of the waters onto its territory via the Bhagirathi channel. Departing from the basis of hydraulic investment, in principle entrusted to the Indian states, the Indian federation took charge of the project, based on the reasoning that the watercourse and the large port involved both states. The division of Pakistan and the creation of Bangladesh in 1971 played the cards into the hands of India as opposed to its weaker neighbor.

1 Embassy of the Kingdom of the Netherlands, 2014, cited by [RAM 18].

The critical period of the year for the management of river waters is during the low-water months, May and June, when the Ganges flow rate dips below 1,560 m^3/s, a rate from which India reserved 1,250 m^3/s for the port of Kolkata and the Hooghly! To find the water required to satisfy both parties, the construction of reservoirs in Nepal had Bangladesh's approval, whereas India preferred to divert the waters of the Brahmaputra towards the Ganges during the dry season. Things stayed as they were; then, India imposed a trial period on the diversion of water from the Ganges into the Bhagirathi–Hooghly, during a period in which the hydrology of the Ganges was going through a long period of drought (1975–1988). In practice, Bangladesh only received 790 m^3/s during the dry season and, in 1977, it had to agree to sign a treaty with India, the Ganges Water Agreement, without international support behind it. The policy of leaving no choice in the matter benefitted India: each of the two countries has 50% of water if the flow rate of the Ganges at Farakka dips to 1,980 m^3/s, and a simple bilateral consultation is planned if the flow rate falls below 1,400 m^3/s, in that case with no guarantee for Bangladesh. The situation is critical, but relations between the two countries are considered to be better than they were between India and Pakistan before 1971, which should create a more favorable diplomatic context[2].

The issue of sharing low water levels is paired with the other, perhaps just as serious, issue of pollution of waters in the Indian part of the Ganges. The river receives 1.6 km^3 of polluting effluents each year (from the chemical industry, treatment of leather and metals and the remains of incineration); in addition to this, there is bacterial pollution, including *Escherichia coli*, with input of fine sediments from different sources, etc. India is making real efforts to reduce the pollution generated by nearby activities, and these benefit from significant international aid, in particular from Japan and the Netherlands. Nevertheless, they remain highly insufficient in the eyes of many observers. Moreover, low flow rates encourage intrusion by salt water into underground reservoirs closer to the surface, infringing on its potability and on land irrigation. The deteriorated state of surface waters and underground waters has led to investment in extracting underground water from deep reservoirs, but a new issue has come up: pollution by arsenic that the sediments naturally contain, which constitutes one of the most serious threats to public health.

3.1.2.7. *A frightening upstream domination of the Brahmaputra*

It should be noted that the sedimentary budget of the Brahmaputra and the watercourses descending from the Indo-Burmese mountain range is unfavorably affected by human activities due to deforestation and bad agricultural practices. Fluvial channels are filled in by excess inputs and therefore lose part of their

2 From the Bengali point of view, see [BEG 87]; see also [SOO 11] for a point of view from outside of the conflict.

capacity to evacuate floodwaters. However, this sedimentary surplus may only be temporary. It is expected to become a deficit, and perhaps a significant one, when the envisaged reservoir-dams are commissioned.

Indeed, what will the future hold, knowing that China and India are constructing reservoir-dams on the upper Brahmaputra? The upper part of the river, or the Yarlung Tsangpo in its Tibetan reach, is 1,625 km long (part of a total length of 2,880 km). The river and the tributaries in its upper basin have considerable potential, 79,000 and 114,000 MW, respectively. A chain of 28 reservoir-dams is planned. The first dam, the Zangmu, was completed in 2016 (510 MW); this is a gravity dam constructed at an altitude of 3,300 m. The dams are supposedly inoffensive because they are classified by China as "run-of-the-river" dams and they have no holding capacity. The Zangmu dam, 116 m high, does, however, hold a volume of water of 86 hm^3 and we do not know if it is equipped for flushing the sediments that are necessarily going to be stored in the reservoir [BUC 14]. Long underground diversions are planned, as well as turbine operation with large daily flow rate variations.

The Chinese dam, which could have the greatest impact, is the Great Bend or the Shuomatan Point, near Medog, where the river crosses the Himalayan mountain range to reach the plain. This dam has an installed power of 38,000 MW, a productivity of 320 TWh/year; this is more than 20% of all Chinese production in the year 2000 (double that of the Three Gorges Dam). This section has the greatest potential in the world due to its height change (2,400 m) and its flow rate. Hydroelectric development will be carried out for the benefit of Tibet and above all China, which will consume the energy produced. If China implements its projects, the consequences for the Brahmaputra would be an increase in its low-water flow rate due to the release of monsoon rain water and snow melting from storage, as well as an alteration of sedimentary flow rates.

In the long term, the Chinese have another project, namely to divert part of the flow rate of the Brahmaputra to the arid provinces in the north, the Xinjiang and the Gansu, all with the end result of benefitting the Huang-He. The Nagong–Huang-He canal is likely to traverse the Salween, the Mekong and the Yangtze. There are enormous challenges, because this involves using nuclear energy in civil engineering and construction work in one of the most seismically active zones in the world. The flow would pass through turbines along the way to compensate in part for the high costs of elevating water to traverse mountainous areas. Since the monsoon gives 70% of its flow rate to the Brahmaputra on the south face of the Himalayan chain, the diversion will only have a minor effect on the flow rate during high waters. On the other hand, the impact downstream will be very serious in the low-water season, because it is not impossible for water diversion to be nearly 1,200 m^3/s in order for a reservoir-dam to be constructed on the Nagong (a tributary of the Yarlung Tsangpo). This will be just as worrying for the Meghna Delta, where a fluvial flow rate of 2,000–3,000 m^3/s is required to halt the salt wedge.

The competition is serious, to the point where a water war between India, as the dominated country with regard to this question, and China is not excluded by some analysts, but potentially Bangladesh bears the greatest consequences. China's preference (in the same way as India) for bilateral agreements at the cost of multilateral consultations does not bode well for the delta's future and its inhabitants [GRU 13, RAM 15]. Then, India reinstated its policy of large dams with the aim of irrigating land and balancing carbon emissions, which are growing rapidly in parallel with the considerable energy requirements. The southern slopes of the Himalayas are the last frontier for hydroelectricity, where 28 of the 32 rivers will have installations, which will reach one of the highest densities of installations in the world. The presence, and threatened existence, of Bangladesh downstream does not prevent India from constructing dams on the Brahmaputra. The same is true for Dibang in the state of Arunachal Pradesh. It will have a height of 288 m and a power of 3,000 MW, but its construction has been postponed since 2008 for environmental reasons. The long-term risk is of course of trapping sediments from the Himalayas in reservoirs and thus of reducing the solid flow rate that supplies the Bengal delta; an increase in salinity of the soils is another major risk. India's objective is to divert part of the flow rate towards the Ganges, which could allow the level to rise during the spring low-water period on the Meghna (from 5,250 m^3/s to 5,850 m^3/s) and be favorable to water management in the delta [BRI 08]. However, India is also planning a transfer from Dibang to the Mahanadi (east-central India) to irrigate the dry lands by skirting around Bangladesh. While the Brahmaputra–Ganges link is supposed to maintain water in the basin and negotiations between the two parties could result in an agreement in the best interests of each party, the second transfer would be carried out with detrimental consequences for the resources in the basin and as a last resort for Bangladesh. But India also plans to block the upper Meghna at Tipaimukh without making any reference to its neighbor. The threat to Bangladesh from large hydraulic works would then be more Indian than Chinese.

Bangladesh's situation therefore contains a truly terrifying element in the medium term. This deltaic country lives in an illusion of equilibrium between subsidence of its area below ground, cyclones and the rise in sea level, all of which are factors that are difficult to compensate for despite the colossal Himalayan sedimentary input. While this unsatisfactory equilibrium was compromised locally by a badly constructed agricultural development in the 1960s, the hope of corrective solutions remains. However, this country and this delta will not resist the processes at work in the Himalayan range. The rivalry between China and India at the beginning of the 21st Century with uncoordinated initiatives to control the strength of watercourses is an illustrative example of the upper domination of large rivers. Alas, the phenomenon is shared with the rest of the Earth's surface, but the singularity remains that in the end, it is going to destroy one of the poorest countries in the world while the latter will not have the least possibility of defending itself. The questions of shortage and pollution are also critical for Bangladesh, in the basin where resources are badly managed and where the international environment is unfavorable to this. Only a coordinated approach at the scale of the basin could help Bangladesh to overcome its difficulties, which are considerable. Being poor and living downstream of fluvial basins traversed by broken political frontiers is no good thing [SHA 09].

3.1.2.8. *Forecasts for global warming from the present moment until 2050: vast human migrations*

In a country which already has a high population (163 million inhabitants) and whose population is going to increase (180–190 million before 2050 is predicted), internal migrations within Bangladesh are going to multiply in number. Internal migration scenarios all produce a figure greater than 13 million migrants, and the majority would be climatic migrants. A 1 m rise in the level of the ocean, combined with cyclones, would cause the country to lose a surface area of more than 4,800 km^2; for a hypothesis of +1 m, this would be more than 12,000 km^2, which is 8% of the area of Bangladesh. The valleys will be affected by more intense monsoon floods, and the country as a whole will be affected by more severe droughts and a reduction in agricultural productivity. Initial migrations could in particular affect the coastal part of the delta and the riverbanks, whereas the migrants would head to the cities of Chittagong and Dhaka. Bangladesh may be the world's non-insular country that is most affected by global warming. A less unequal development and the overall reduction in greenhouse gas emissions alone could reduce this phenomenon [RIG 18].

These forecasts take on particular importance when seen in the perspective of regional political crises that are themselves responsible for forced displacement of people. Persecution of Rohingyas, excluded from Burmese citizenship since 1982 and expelled from Burma, is expressed by a brutal emigration into Bangladesh. The refugees have ended up being collected together by the Bangladesh government in the Thengar Char, an island in the Gulf of Bengal which emerged approximately 10 years ago off the coast of Chittagong. Yet, this island is very low-lying, flooded by the annual floods of the Brahmaputra, by high tides and *a fortiori* by cyclones, and above all has no form of development that would allow the integration of refugees into the country's economy. The conditions for a humanitarian disaster are therefore all present [AZA 17].

3.2. The Mekong Delta in a suspended status

3.2.1. *A technical machine, constantly more complex*

3.2.1.1. *Human generalities*

The Mekong Delta covers between 40,000 and 50,000 km^2 depending on the definition used, primarily in Vietnam. The delta has two water supply sources: rains and the Mekong floods. The rainy season takes place from May to November. While the northern part receives less than 1,500 mm/year, the Ca Mau peninsula receives more than 2,000 mm. During the dry season, the soils in the upper delta dry out, but those in the lower delta maintain a certain level of humidity. The monsoon river flood blocks the evacuation of rainwater that accumulated in the Transbassac from the month of August onwards and the rest of the delta from the month of September onwards.

The delta houses and feeds 18 million inhabitants, which gives it a population density of 440 inhabitants/km^2. Around the year 2050, the population should reach 30 million inhabitants, the population density should reach 750 inhabitants/km^2 and the area dedicated to agriculture should recede to 0.083 ha/person. The degree of urbanization, which is approximately 20%, should reach 50% in 2050. The delta guarantees security of life and food for Vietnam and Cambodia; with its 2.5 million hectares that guarantee three annual harvests, it is also considered as the "rice bowl" of Southeast Asia and makes Vietnam the second- or third-largest exporter in the world of rice at competitive prices. Agriculture in the delta produces 40% of the total value produced by Vietnam; it contains fishing grounds, an area of aquaculture which is among the most productive in the world, and ecosystems of very high value. Contrary to a delta like the Mississippi, which guarantees tangible food production, but which is not essential for the regional population, the Mekong Delta takes full advantage of its annual flood and fertile sedimentary input to benefit human food supplies (Figure 3.2).

3.2.1.2. *Formation of the delta*

The formation of the Mekong "megadelta" began between 8,000 and 6,000 years BP, at the end of the period of rising sea levels, by estuarian aggradation deposits which have filled in a valley that was incised in the quaternary deposits when the ocean level was at its lowest. Once the sea level stabilized, at approximately 6,000 BP, the sediments prograded out of the estuary and encroached on the ocean. At approximately 3,000 years BP, the model changed from being a "tide-dominated" type, featuring mangroves, to a mixed "tide-dominated and wave-dominated" type, whose energy explains the formation of a sandy coastline with dunes, since this is influenced by monsoon winds from the northeast [TA 05].

Following the stabilization of the sea level, the river mouths progressed on average by 16–26 m/year in the East China Sea. Their prominent morphology has exposed them to the attacking forces of waves and tides, as demonstrated by the series of sandy coastal bars lying parallel to the coast (tides can reach an amplitude of more than 3 m off the coast of the mouths). However, the delta has also progressed significantly southwards in the sheltered zone where mud originating from the river is deposited, allowing the growth of large mangroves [ANT 15]. The Mekong Delta is thus dominated by the trio of influences from the river that constructs it, the action of tides in the East China Sea and the Gulf of Thailand, and waves. The relative proportion of these influences in the dynamic of the coastline varies depending on the sectors of coastline (direction of waves, tide amplitude and angle of attack of the waves). At the apex of the delta, the river has very low energy, just capable of carrying sand along its bed, whereas its suspended load in principle enriches the soils in the deltaic plain each year. In actual fact, if the alluvial plain in Cambodia traps 10–20% of the suspended load, the deltaic plain traps sediments over a much wider area.

Figure 3.2. *The extent of flooding (in blue) in the Mekong Delta. The flood peak has passed and the flooded area is receding at the end of the monsoon period (November 2011). (Source: Satelligence/NAAEOSDIS). For a color version of this figure, see www.iste.co.uk/bravard/sedimentary2.zip*

The immense alluvial construction constitutes the delta *sensu lato*, but the term estuary is frequently used for each fluvial branch of the delta. The reason is that the continental mass created by alluvial build-up during the Holocene is considered to be separate from the channels which form a multiple estuary. The arms, over-incised by high flow rates during the monsoon (20,000–40,000 m^3/s in October), are approximately 10 meters deep at a distance of between 30 and 140 km from the sea. The effects of the dynamic tide in the estuaries are felt far upstream during the dry season, reaching up to Phnom Penh in certain years. Each estuary undergoes a twice-daily amplitude of 3.2 m. The depths are shallower (only 5 m) in the lower part of the channels.

3.2.1.3. *Technological choices and economic growth*

The Mekong Delta was previously populated by the Khmers, who dug canals as early as the 3rd Century AD around the large port of Oc Eo, which depended on the Funan Kingdom (1–7th Centuries). In the 17th Century, hydraulic development set up for exchanges, and to protect and feed the population, was part of the Precolonial Vietnamese traditions of the Red River Delta, motivated by the rulers who were anxious to avoid famines and social problems. A wave of Vietnamese migration from the basin of the Red River then occurred in the Delta; this was the "southward advance". The Khmer people had been displaced to the west of the Bassac into the Mien Tay, whereas the Vietnamese and Chinese had settled in the Mien Trung, between the Mekong and the Ho Chi Minh City (Saigon) region. Under the authority of the Nguyen dynasty, infiltration of the Vietnamese was accompanied by the digging of the initial channels and the development of rice cultivation. The network of waterways was extended for the benefit of French colonization at the end of the 1880s, under the guidance of the public works department and with a decisive contribution from the body of engineers, dredging companies and large landowners [BIG 09]. In the 1890s, the upper part of the delta, flooded by freshwater during the summer monsoon, was the country of floating rice. In the lower part, closer to the sea, the Mien Tay encompassed 19,000 km^2 of swampy forests, including mangroves on salty soils in the tide-dominated areas. The current network of waterways was created by the French governorate in order to instill military security in the delta and subsequently to facilitate agricultural development. More than 3,000 km^2 of forests remained in 1938 after intensive clearing under French colonization. In the intermediate zone, which received water from rain and was well suited to the cultivation of rice as its only crop, the governorate developed the network of waterways. It imposed statutory labor while introducing dredging from 1884 onwards, in such a way that in 1930, more than 14,250 km^2 had been drained and made cultivable, benefitting large French concessionaries and small Vietnamese farms. At the end of the 1930s, protective dykes and anti-salt dams were constructed, following the example of the Red River Delta, in a move towards rice cultivation in paddies. From the end of the 1950s to the middle of the 1970s, this same system was repeated in the plain of the Roseau River by the Americans and to the great benefit of public works companies. The war affected the delta in terms of napalm spreading and led to large population movements towards "fortified villages" to flee the conflicts. The unification of the country by the communist government was accompanied by migration towards the delta of a partly Catholic population that originated from the North; development of the delta's resources followed the model of centralized socialist management, despite the resistance to collectivization that was expressed by low-level peasantry. The delta became the "irrigation front" at the service of the nation and was managed by hydraulic engineers from Hanoi despite their lack of knowledge of the local hydraulic conditions. They were no more accustomed than their recent predecessors to the fact that seasonal flooding, both slow and fertilizing, did not present a constraint as it did in the Red River Delta. As a result, errors were made such as the construction of pumping stations, which were in principle suitable for polders, but not for the agricultural outskirts of the Mekong Delta. In the 1990s, the hydraulic network was reorganized in order to take irrigation into account and to resolve specific handicaps, in parallel with the success experienced after the economic liberalization of 1986.

Today, the network of waterways extends to more than 50,000 km; along with the network of dykes intended to provide security for agricultural paddies, it has had a profoundly disturbing effect on water circulation, in such a way that flooding of the delta is cut off from the flooding regime. Isolation of the paddies has had the effect of reducing productivity of the ground and increasing salinity of the soils. Flooding of the plain greatly depends on practices applied by humans, such as the opening and closing of locks controlling the spread of freshwater and salt water in the channels between tides. This contrasts with the continuity and the progressivity that is particular to natural flooding. From this complexity, a pattern of flooding emerges that takes on a highly mosaic quality [HUN 11].

Geopolitical analyses of historical management of the delta have been carried out. Even though coordination has generally failed, the Mekong Delta has always been under the control of technical bureaucracies (the very powerful dykes service), of "hydrocracies" [BEN 14] which are on the same level as the hydraulic companies mentioned in the Wittfogel model [WIT 57]. The initial technical landscape has been perpetuated, constantly enriched, up to the point of possessing a "physical inertia" that is incompatible with all changes to the frame of reference. Maintenance of the "hydro-agricultural" machine, assimilated to an "institutional inertia", has turned out to be extremely costly for the local population, because it has induced considerable maintenance costs, above all increasing costs as the system progressively complexifies (e.g. maintenance of the dykes and the costs of dredging the "humpbacks" created by sedimentation in the channels) [BIG 09]. Under the French governorate, the technical choices have increased twofold with the installation of precarious population service providers who live at the mercy of floods in a frontier zone that is in the process of being cleared; the most vulnerable were in the provinces of Dong Thap and An Giang, which are lower-lying and therefore flooded more; the populations there were more precarious. At the end of the 1960s, the United States focused on the delta by trying to introduce methods of the Tennessee Valley Authority into it under the framework of the Mekong Delta Development Program (MDDP), but without managing to do so while the war was ongoing. These principles, based on hydraulic control of nature, were implemented after the victory of the communists in 1975, demonstrating that since French colonization, the continuity of concepts has transcended political regimes, and water control has taken over from adaptation [KAK 08]. As for the population itself, it has demonstrated an exceptional ability to adapt to difficulties and expressed forms of opposition to all the political and technical regimes by developing methods of flexibility in the face of natural constraints; this results in an individuality seen at times as a threat to the achievement of objectives that demand high levels of coordination and even to alternative, local means of adaptation that are advocated by the authorities [BRO 85].

3.2.1.4. *Symptoms of the crisis that are recorded in the delta*

The initial signs of environmental change in the delta have been observed for approximately the last 25 years and interpreted depending on explanations proposed by researchers, but have not always been accepted by the political authorities; at least, this is the case for results that question recent or current decisions. The changes relate to the deterioration of agricultural resources, the subsidence of the plain aggravated by pumping that is too intense, the damaging role of a canal network that is perhaps too spread out and the rise in sea levels due to global warming and destruction of the mangrove forests.

Pumping, subsidence of the deltaic plain and the threat of submersion by rising sea levels

The most obvious and most precocious sign of deltaic malfunction is subsidence and its accentuation since water pumping increased in the 1960s, to the point of having one million drilled wells to supply more than 20 million inhabitants (30 million according to certain sources, where this uncertainty is related to choices of area perimeters).

In addition, the subsidence is combined with the rise in sea levels. This rise is in the range from 6 to 15 mm/year depending on the coastal sectors. In a specific national program, Vietnam anticipates the rise in sea level as a function of two scenarios from today until the year 2100, one at +75 cm and the other at +100 cm. In its natural state, the rise in sea levels should be distinguished from the intrusion of salt water.

Overexploitation of water has led to an increase in subsidence [ERB 14] due to compaction of the levels that have been drained by exploitation of underground waters. Reduction in the level of the top of the water table is most significant in the sectors of Ca Mau southwest of the delta (maximum of 70 cm/year) and in the area contained between the Mekong and the Bassac (70 cm/year), and it is also significant in Ho Chi Minh City (according to a central band on a NE–SW axis). Overpumping, due to the digging of 100,000 wells, induces a rate of subsidence ranging between 0.3 and 3 cm/year for the entire area of the delta, with a record 4 cm/year in Ho Chi Minh City. In order to understand the significance of these values, we need to remember that the absolute speed of the rise in sea levels is from 0.2 to 0.4 cm/year, which is approximately 10 times less. By extrapolating these values, specialists model a reduction in the elevation of the ground surface and an additional submersion height in the range of 0.4–1.6 m before 2050. The spatial irregularity of subsidence and of its intensity is an indirect effect of overpumping and of the damage of surface infrastructures. To complete the picture, there is an increased risk of intrusion of salt water of marine origin into the reduced underground freshwater reservoirs and a risk of contamination by arsenic present in the waters pumped from great depths.

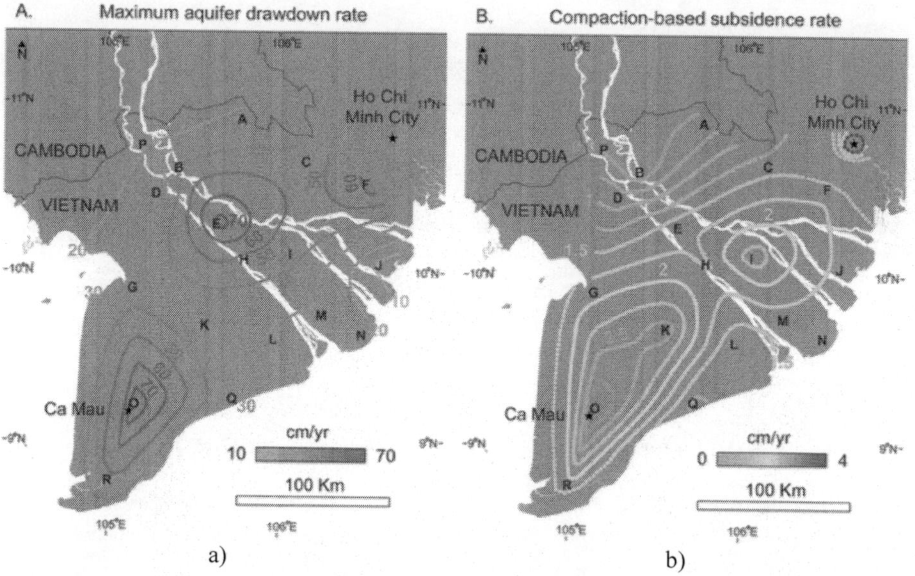

Figure 3.3. *Dynamics of the underground reservoirs and of subsidence in the Mekong Delta for the period 2006–2010. (a) Reduction in height of the aquifer (cm/year). (b) Subsidence of the soil (cm/year) measured by radar imaging [ERB 14]. For a color version of this figure, see www.iste.co.uk/bravard/sedimentary2.zip*

The rise in sea level and intrusion of salt water

Considering the problem from the point of view of its effects, one aspect of this question is the effect of salt water on the population: it is recognized that the threshold of 2–3 g/L of salt must not be exceeded in drinking water; yet the level of 4 g/L is exceeded locally within the delta. How can this situation be explained? The delta opens out onto the ocean via 17 estuaries with a total width of 25 km along an 850 km seafront, which facilitated drainage of the land, but the very long coastal façade leads to considerable exposure of the inside of the delta to inflow from the sea. The dynamic tide reaches 350 km upstream during the dry season, up to Phnom Penh, and from 150 to 200 km during the season of high waters. The intrusion of salt, associated with the dynamic tide, is spread over 50 km in the delta's estuaries during the dry season (the total discharge level between 1,700 and 5,000 m^3/s) compared to that over 20 km during the high-water season; it has advanced by 20 km over the last 20 years. As a result, there is an exchange of water that reaches 1.5 or 2 km^3 each day during the dry season. It can be feared that the rise in sea levels increases the energy of tides and the intrusion of water over time, as well as sedimentation in canals, and that it obstructs drainage of the rice paddies. The cities are also affected; more than 30% of the area of Ho Chi Minh City was affected by flooding and Can Tho by more than 50% [HOC 12].

Another manifestation of change is the intrusion of salinity from the sea in favor of a very dense network of artificial canals. The saline intrusion affects the chemistry of water over two million hectares as well as the quality of water destined for consumption by the population; it also reduces the beneficial effects of freshwater from the Mekong, which is necessary for rice cultivation. Penetration of salt via the channels at high tide has already seriously affected rice production during the dry season and reduces the production to a single harvest during the rainy season.

In response to deterioration of the quality of agricultural waters by salt, the Vietnamese authorities started work in 1994 with the aim of protecting 160,000 ha. Canals have been dug to carry water from the Mekong at high water, and anti-salt locks have been constructed on the Ca Mau peninsula to block off the sea entrances at high tide. This procedure has allowed rice production near the Mekong to be increased, thanks to a move towards two or three harvests per year.

During the wet season, floods will continue to affect the upper part of the delta (over 15% of its area), but will be confronted by the rise in sea levels, in such a way that 60% of the delta, located in the lower part, would suffer from increased vulnerability. The complexity of maintaining rice cultivation in the face of rising sea levels will be a combat that has led to rice being considered as "frontier cultivation"; however, the rise in water levels should be sufficiently slow for the construction of protection structures and the regulation of waters to be envisaged stage by stage and at a reasonable cost [WAS 04].

Pollution of potable water by arsenic

Lastly, and this is extremely concerning, underground waters in the Cambodian and Vietnamese part of the delta, pumped at depth from recent alluvial deposits (Pleistocene and Holocene sediments), are enriched by natural arsenic, adsorbed* by mineral particles containing iron in the form of oxides and placed in solution in a highly reductive environment. Waters at shallow depths, contaminated by fecal bacteria, are no longer consumed; deeper waters have been used in their place since the end of the 1990s. These waters in fact have a concentration of arsenic that is dangerous for human health[3]. Between 0.5 and one million people are affected in the Mekong Delta, much less, it must be said, than in the Red River Delta (10 million people) and in Bangladesh. Treatments by filtration are available to significantly reduce the level of arsenic in the water consumed [BER 07].

However, the significant amount of dyking and the increasing control of water in basins that are disconnected from the fluvial system, along with introductions of sea water, have led to ecological disorders such as a reduction in biodiversity, silting-up of channels and erosion of the banks. Since the decade 2000–2010, Vietnam has chosen to reinforce sea protection infrastructures to protect itself from cyclones and to anticipate the predictable effects of climate change.

The situation is uncertain for low-lying areas. One of the negative effects of the channels and of the locks was lowering of the underground waters and oxidation of the natural sulfur content of the soils, which liberates sulfuric acid; leaching by rains in season leads to acidification of the waters in the channel (pH values below 3 are possible) [TUO 03]. This brought on a cascade of consequences: the water is not easily drinkable, fishing in the channels and production of farmed shrimp in briny water has decreased due to the effect of lower introduced quantities of briny water, and the acidification of waters. This negative spiral, due to the partial comprehension of the impacts of developments, has in particular detrimental effects on the communities of poor farmers and agricultural workers who require vegetable proteins from rice and animal proteins for food and even for basic survival. Lastly, the surface sediments contain iron in the form of sulfides.

Sedimentary inputs and coastal erosion

It is recognized that growth of the deltaic front has been continuous over the last seven millennia and that it has taken place at an average rate of 30 m/year; it has been calculated that the load has been more or less constant over this period of time, due

3 Exposure for 10–15 years to chronic concentrations of 50 micrograms/L leads to poisoning that manifests itself as diseases and cancers of the skin, the lungs and the liver; the nervous and vascular systems can be affected. Europe and the World Health Organization recommend a standard of 10 micrograms/L.

to a lack of other arguments [TA 02]. A study based on spectrometry of the color of the ocean and measurements made in the sea has been carried out for the period 2003–2012; it analyzes seasonal concentrations of particulate matter in suspension in the plume located at the mouth of the estuaries. The study demonstrates that the input from the Mekong in flood has diminished by 5% on average per year over this period, that the plume in the sea has been reduced and that in winter the tide and the waves have the strength to erode and drag sediments from the prodelta along the coast in a southwesterly direction [LOI 14] (Figure 3.3). Another study, carried out on the basis of 43 years of monitoring by the Landsat satellite (1973–2015), has allowed evaluation to have a little more depth in terms of time, and a more complete overview of the river's path, in comparison with input from recent measurements. Whereas the delta has prograded by 7.2 m per year over the period 1973–1995, and by 2.8 m per year between 1995 and 2005, it has receded by 1.4 m per year over the period 2005–2015. The length of the coastline undergoing erosion has exceeded the length that has prograded since 1979, and the surface budget has also been receding over the period 2005–2015 [LI 17].

Figure 3.4. *Coastal erosion destroyed the mangrove forest and aquaculture enclosures. (Source: J.-P. Bravard). For a color version of this figure, see www.iste.co.uk/bravard/sedimentary2.zip*

The coastal and estuarian processes, in the same way as the set of two tides that affect the delta, will without doubt be accentuated. The delta is likely to evolve towards greater wave action, and it is probable that the reduction of a sandy input (refer to *infra*) is going to play out along the same lines, while the sediment deficit at the mouths can be expressed by a more rapid recession of the coastline.

3.2.2. *Extremely worrying emerging factors*

Vietnam is justifiably fearful of the need to simultaneously manage fluvial processes and coastal processes, of which still very little is known about the basic outlines.

3.2.2.1. *Expansion of irrigation to Vietnam and Cambodia: a threat to low-water flow rates and underground reservoirs*

The success of Vietnamese rice cultivation is due to perennial irrigation for 60% of the area in the Mekong Delta, which leads to production all year round and provides half of national production. Cambodia is still a rice producer that relies on rains in the wet season; in order to improve its food security and, just like its neighbor, draw revenue from rice exports, it is making efforts to massively increase its production. Rich agriculturists require more surface water and, since the middle of the 1990s, they have been turning increasingly towards motorized pumping of underground water to be able to produce a second harvest during the dry season (the requirement for water from pumping is nearly 600 mm, taking into account the rains, the evapotranspiration and the losses due to infiltration). Although predictions in terms of extraction are considerable for upstream countries, the downstream countries are facing worrying perspectives in terms of extraction for domestic use and above all agricultural use, with the population undergoing rapid growth, in particular in Cambodia. Vietnam fears water transfer both inside the basin and outside the Mekong basin. This is already the case in Thailand, and could be the case in China if the waters of the Nu and the Lancang were to one day supply a south–north transfer via the Yangtze, as projects in the 1980s had proposed; it is more probable that some of the waters of the Lancang will be deviated in the short term to benefit agriculture in the Yunnan [TRA 10].

Longer-term fears relate to two major factors, namely the reduction of flow rates during periods of shortage and an effect on the resource of underground reservoirs. One the one hand, extraction from the Mekong has increased, which results in a reduction by 30–40% of the low-water flow rate in the river and the Bassac, and then in a reduction of the flow rates available in the Vietnamese channels of the delta. One the other hand, the level of underground reservoirs has been reduced, in which the waters are to a large extent fossilized and do not replenish, and whose salinity increases. The reserves certainly allow the situation to be maintained for several centuries, but the difficulties reside in the fact that domestic pumping of pure water at low cost will become obsolete, and then that arsenic found at great depths will rise to the surface and that subsidence will affect the border region [ERB 16].

In other terms, the development of rice irrigation is not sustainable at this pace and with these choices, without even taking into account the satisfaction of similar needs that are expressed further upstream. It would be preferable to extract smaller volumes for use on high value-added crops and try to spare smaller agriculturalists, or even to make a transition from agriculture to aquaculture when salinity requires it.

3.2.2.2. *Construction of reservoir-dams in the basin and their effects on the Mekong Delta*

Regulation of hydrology of the Mekong is already significant because Chinese reservoirs that have been constructed or that are in the process of being constructed on the Lancang (Volume 1, Chapter 5) have the ability to retain part of the monsoon floods and then release water during the dry season to support the production of electric current. The dams planned on the tributaries will have a similar impact to those of the Lancang range. On the other hand, dams planned for the future on the main channel of the lower Mekong belong to the run-of-the-river category, without necessitating annual regulation. The primary consequence already relates to the reduction of high flow rates during the wet season. The volumes stored in the reservoirs are limited, but they could delay the start of high waters by two weeks.

Regulation will in theory not just have negative effects. China correctly asserts that the increase in flow rates of low waters will be beneficial to navigation and to irrigators downstream, but we can suppose that the consumption of this water will be so great upstream of the delta that any possible positive effect of releases will be cancelled out or very much reduced by the additional consumption. Inundation by the annual floods will no longer be of the same height, the same duration or the same spatial extent, particularly since the protection dykes are increasing in number. An indirect consequence of this is that leaching of acidic sulfates that have accumulated in the soils by evapotranspiration during the dry season will no longer be guaranteed to the same level of efficiency, at risk of depleting soils and making yields decrease [PHO 95]. Another study modeled a flood with a return period of 20 years, of the kind seen in 2000, whose effects have been aggravated downstream from the dyked sections; the rise in sea levels will increase the height of floodwaters by 2.5 m over an area that can reach more than 4,000 km^2. The dykes currently in place prevent drying-out of flooded land whose waters are acidic, block entry of suspended sediments to irrigated paddies and encumber the estuaries with fine sediments [HOA 07].

However, taking into account the options selected for modeling of global warming and consumption scenarios at the scale of the basin when it is enlarged to include regions of transport, the monsoon floods are expected to increase before 2050–2060 and the dry season low waters to reduce even further. Modeling that anticipates the construction of dams and rising sea levels highlights the primary position of responsibility associated with placing installations in the watershed instead of the rise in sea levels or certain global warming effects such as the increase in flood peaks. Flooding would increase slightly and the sedimentary inputs would reduce by 40%, in such a way that the production of rice, which currently depends to a degree of 50% on the fluvial input of nutrients, would fall to negligible values if the efficiency of entrapment in reservoirs was itself taken to a value of approximately 50%; fishing productivity would fall for the same reason and due to the dams' situation as obstacles to migration [MAN 15].

Concerning sedimentary transit, the hydroelectric development of the Lancang "cascade" is of primary importance, since the upper Mekong drained 23% of the watershed and provided 15% of the annual flow rate, but guaranteed 65% of the annual suspended load. Studies carried out consider that the reservoirs of the upper Chinese basin currently trap between 80 and 90% of the sediments that were measured before they were built (at Chiang Saen, at the entrance to the Lower Mekong), and still, 50% reach the delta due to inputs from the tributaries. However, there is an obvious disproportion between the role of the high mountains, even though they receive little rain, and that of the mountains and hills in the Lower Mekong basin, which receive a lot of rain but which are not very productive, in particular in terms of this sand that has allowed the delta to continue growing for several millennia. The existing dams are roughly 15–18% efficient at trapping the load that they currently receive, and it is essentially a case of reservoir-dams constructed on the Lancang. If the eight reservoir-dams planned for the Mekong were constructed, the entrapment efficiency would be approximately 80%, but it would reach approximately 90% if all the dams planned, meaning that more than 130 at the scale of the basin, were constructed [KUM 10]. Another evaluation cites the entrapment of 50% for the existing dams; this figure even reaches 96% for the planned dams, which means that only 4% of sediments would reach the sea once the stocks in the Mekong channel have been exhausted. The specialists recommend equipping the dams with gates at bed level to "flush" the sediments through and to outfit the existing dams *a posteriori* in order to improve the situation downstream and to lengthen the lifetime of the reservoirs [KON 14]. This optimistic scenario simultaneously assumed that extractions would stop completely, which is not a very credible option in light of current perspectives.

By way of an example, serious concerns relate in a first instance to a reduction in infilling of the Tonlé Sap by the Mekong flood. The time taken to fill it will be shortened by approximately 15 days, and the relative height will be reduced if the beds of the Mekong channel continue to diminish at Phnom Penh, as was occurring as a direct consequence of the extractions. The inputs of fine sediments and nutrients are no longer expected to allow the river to guarantee the same planktonic and fishing productivity as in the past (Figure 3.4). The increasing population that makes a living from fishing in the waters of the Tonlé Sap and from recession agriculture on the banks will suffer significant consequences. Incision of the bed by extractions that take place in the Mekong channel already accentuates this phenomenon.

3.2.2.3. *Overexploitation of fluvial sand*

The exploitation of sand and gravels, when they exist, has direct negative effects on the sedimentary fractions that are useful for maintaining the deltaic coast. In the specific case of the Mekong, it is all the more delicate to admit that the actual sand part was ignored until a very recent date. The paradox is that the sandy component of solid transport has not been measured on the river since 1960, but this has not worried those in charge because the sand was highly visible on the banks throughout the fluvial path and it arrived at the coast. Since the sand arrived from upstream and it was very easily accessible, anyone and everyone has been able to dredge it from the channel bed and this has always taken place (Figure 3.5).

Figure 3.5. *The floating village of fishermen at Chong Khneas to the northwest of Tonlé Sap during the dry season. (Source: Google/Digital Globe, image taken on April 24, 2004). For a color version of this figure, see www.iste.co.uk/bravard/sedimentary2.zip*

The river has sunk locally, as demonstrated by the level of the waterline at low waters; deep pits intersect the delta channels, and the extractions have reached deep ancient sandy deposits; the banks collapse, carrying away roads, dykes and chunks of port infrastructures. Ignoring this factual information, recession of the banks of the Mekong has mainly been attributed to climate change, which is rendered responsible for flooding due to the ENSO (El Niño Southern Oscillation), the oscillations of which reinforce the monsoon [DAR 13]. It is dangerous to absolve societies from their mistakes by attributing responsibility for these effects to global warming, which is only one factor among many others. The impacts have propagated along the coastline. The fact that the arrival of sand at the estuary mouths is very much reduced by reservoirs and extractions and that they no longer supply the coast is a reality that must be taken into account, and which turns out to be expensive. Selective extraction of sand certainly has the effect of increasing the relative part of mud in the river and the branches of the Mekong estuary, like in other tropical regions, which affects the equilibrium and the ecology of the coasts:

> "In most of the tropical world, rivers and estuaries are turning into mud drains as a result of human activities. The adjoining coast and coastal marine resources are also degraded by mud" [WOL 00].

Figure 3.6. *Extraction of sand on a bar in a convexity of the Mekong, displaying dune morphology (downstream of Vientiane, Cambodia). A dredger is extracting gravel from the riverbed in a basin linked to the river; mechanical diggers extract the surface sand. (Source: J.-P. Bravard). For a color version of this figure, see www.iste.co.uk/bravard/sedimentary2.zip*

3.2.2.4. *Effect of a sedimentary interruption on deltaic mouths*

With regard to the delta, one concern of researchers relates to the behavior of estuaries in response to the fact that dams in the basin will impose regulation. Sedimentary transport models predict significant clogging of the mouths by fine and flocculated silts which are (usually) associated with the salt wedge. Due to an insufficient fluvial flow rate, the salt wedge will move further upstream than it does today. Water usage will be disrupted by this.

More precisely, what are the mechanisms at work here? Pedological erosion is intensified in many tropical basins where the management of cleared land is deficient; this is very probably the case for the Mekong basin. On the one hand, agricultural practices increase the quantities of fine materials that come from soil erosion, but, on the other hand, reservoirs retain a proportion of this and it is not possible to establish a credible budget. In this context, where the sedimentary budget is still highly uncertain due to a lack of measurements, the key point appears to be elsewhere. Two factors play an important role.

– The downstream flow of estuarian branches has a very high turbidity, due to the abundance of clays and silts which come from the erosion of the soils in the basin and erosion of the coastal zone. Seasonal regularization of the Mekong flow rate by reservoir-dams is responsible for an increase in the input of mud into the estuary, because the pumping effect of the tide, which carries away the mud at the cost of the prodelta, will no longer be contradicted by the flushing effect of the fluvial flood. This factor alone could account for 17 Mt/year according to one model; the same study anticipates entrapment of heavy metals in muddy sediments [WOL 00]. The quantity of mud is such that it causes multiple problems for the managers, in particular through chemical binding of pollutants. The extension of the mangrove would, however, allow "pollution" due to excess mud to be reduced.

– With the same effect, diversions of fluvial water for irrigation, at the scale of the basin (see *infra*), will increase salinity closer to the mouths, therefore having a positive effect on the primary factor of flocculation*.

3.2.2.5. *Threats to the mangroves*

Alas, extreme events are part of the deltaic lot. Admittedly, they are exacerbated by natural events. Other than the Mekong floods (in 2000, human loss of life numbered 500), typhoons have caused considerable human and material losses: typhoon Linda in 1997, typhoon Durian in 2006 (86,000 houses lost their roofs, 20,000 were destroyed and 8,000 ha of agricultural lands were destroyed). It is generally recognized that the fragility of coasts in the face of typhoons and actions of the sea has increased in recent decades, in particular due to recession of the mangroves. Mangroves are a forest formation, essential in order for the coastline to be maintained in the face of rising sea levels and typhoons (Figure 3.7); it is the first line of defense, formed from colonization by various species adapted to the tidal habitat (*Avicennia*, *Sonneratia* and *Rhizophora*) and formerly freshwater species (*Melaleuca*). The area of mangroves with the species *Avicennia* was 100,000 ha and that with the species *Melaleuca* was 190,000 ha at the beginning of the 1950s, on the peninsula of Ca Mau alone, but recession is significant.

Figure 3.7. *A young flooded mangrove forest at the edge of the road to Ben Tri. (Source: J.-P. Bravard). For a color version of this figure, see www.iste.co.uk/bravard/sedimentary2.zip*

By way of an example, the wooded area was reduced by 75% between 1968 and 2003 in a control sector of the Ca Mau peninsula. The initial recession took place during the Vietnam war, following the use of Agent Orange and herbicides (more than 120,000 ha were affected, without taking serious effects on human health into account); after the war, recession was observed due to the rise in rice cultivation, before this receded in turn to give way to shrimp farming (developed areas of paddies cover 40% of the area that was once mangrove forest, compared to 60% of rice cultivation) (Figure 3.8).

This type of farming has indeed undergone a significant boom, and peasants, free to try to improve their living situation, have joined in with the government's plan in favor of shrimp farming, which was launched at the end of the 1980s. Other than the direct effects on the forest area, this commercial activity has potentially damaging consequences in terms of water pollution and the salinization of soils and the underground reservoirs, since shrimps require briny water; in terms of pathogenesis which affects shrimps; then in terms of the reduction of productivity, linked to exhaustion of the organic soils on which the mangrove was rooted; and lastly in terms of loss of revenue for farmers. The system thus described is considered to certainly become a "lose–lose" situation after a short number of years [BIN 05, FRO 02].

Figure 3.8. Shrimp-farming ponds that have been set up in the cleared mangrove, supplied with briny water. (Source: J.-P. Bravard). For a color version of this figure, see www.iste.co.uk/bravard/sedimentary2.zip

The recession of the mangrove ceased in terms of an absolute value in the middle of the decade 2000–2010, thanks to reforestation efforts, but the mangrove is increasingly fragmented, young and deprived of primary characteristics. WWF-Vietnam launched the pilot operation "biological shield" against the sea in the Ben Tri province to restore the mangrove forest, its natural biodiversity (fish, crabs, shrimps and birds) and the services it provides.

3.2.3. *What will be the management choices in the future? Giving preference to the scale of the basin*[4]

The official Vietnamese point of view advocates the installation of a mechanism of truly sharing hydric resources between the six basins in the Mekong, in a "spirit of cooperation for co-development". In particular, this occurs by means of sharing hydrological data and rules of operation of the cascade of hydroelectric structures; this also comes, for all hydraulic projects emanating from a country in the basin, from carrying out environmental impact studies which relate to its effects at the scale of the basin and at several time scales (in the short, medium and long terms). In other words, the Lancang-Mekong cooperation, which was initiated in 2016 by China (Volume 1, Chapter 5), must be based on the principle that the waters of the Mekong present an advantage common to all the countries in the basin. It must set out a treaty based on the principle of what was the 1995 Mekong Agreement and the 1997 United Nations Convention about "uses unrelated to navigation of international watercourses"; the principle of governance of the Rhine is another model [NGU 16].

The Vietnamese government promises:

1) detection, monitoring and evaluation of the impacts of upstream exploitation projects to indicate their danger and negotiate with the governments of upstream countries;

2) revision of the socioeconomic master plan and local plans to conserve potable water, to efficiently use salt water and to learn to live with flooding and drought;

3) changing of the development model, by giving priority to cooperation between the central power and the local authorities, to vertical relations and to education.

At the end of the year 2017, the Vietnamese government announced Resolution 120, which organizes sustainable development of the delta with a view to the effects of climate change.

Lastly, the principle of solidarity in a basin, no matter how much this may resemble a utopia in the case of the Mekong, presumes that the economic costs of socioeconomic and environmental damage should be taken into account in the cost of the hydroelectric structures that have been built and that are planned, that are seen and that are going to continue to be exhibited by the delta. It is obvious that the price of electricity, which is envisaged to be very low, would increase [WOL 00].

4 Sustainable solutions at the scale of the delta are detailed at the end of Chapter 5.

3.2.3.1. *Analysis of the complexity of responses at the heart of the delta*

Work to summarize the situation has presented a critical state of affairs in a system of dynamic production [KAK 08]. The policy of the years 1980–1990, based on exclusive management of freshwater for the benefit of rice cultivation, has been very expensive and has had harmful effects (increased pressure on resources during the dry season, water pollution, lack of fertilization by the floods and downstream effect of the dykes). Today, this policy must increasingly give way to briny water, which is required by shrimp farming and which enters into conflict with the policy of freshwater, in places where rice cultivation is maintained; this conflict would take place to the detriment of the poorest fraction of the population of the delta in the regions where aquaculture requires significant investments. This evolution towards new forms of high productivity is carried out at ever-increasing social and environmental costs. Part of the rural population has migrated towards the cities (mainly Ho Chi Minh City) to work as unqualified labor in restaurants or factories, often without an authorized residency (the population movements are highly regulated by the authorities in Vietnam), which makes their condition even more unstable.

3.2.3.2. *The risk of a single explanatory factor: climate change*

The changes that affect the Mekong Delta are highly complex, among those that are recorded in the watershed (dams, consumption of hydric resources by irrigation, extractions from the riverbed, etc.), in the delta (natural subsidence and subsidence linked to overpumping, drainage, dykes, etc.) and on the coast (rise in ocean level, aggravation of cyclones, sediment deficit, deterioration of the state of the mangroves, etc.). The complexity of the causes is all the more delicate to present to managers and make them understand, since it has a high time component and a certain spatial variability in dynamic contexts.

The situation is so complex that there is a real risk of simplification in order to spread understanding and to explain the social difficulties, in particular migration from the delta towards the cities, by an encapsulating factor which is climate change. Over the last 10 years, the annual net migration for the delta is negative, with the departure of one million inhabitants (– 5% per year), twice the average for Vietnam, in general from the most vulnerable provinces [CHA 18]. Investigating the detail reveals that the causes of migration are attributed to bad harvests, themselves due to salinization of the soils, to exceptional droughts, to coastal erosion and to the construction of dams on the river, with global warming as the primary cause. However, the analysis goes further than this, since part of the migration is even attributed to the impacts of measures taken to protect the population and the harvests from the effects of global warming, where these measures are the construction of dykes that affect fishing and the fertilization of rice paddies by the floodwaters. Thus, global warming would exacerbate the migration in both direct and indirect ways. We do see the danger of an explanation of this kind, which attributes responsibilities at a global scale, including the mistakes made in the management of the delta, which would be an unsuitable response to global warming. These simplifications lead us to run the risk of exonerating those in charge of water management, along with many economic sectors of the basin, from their responsibility and from the requirement to resolve all the problems that they are responsible for in an integrated manner.

Scientific circles are themselves tempted by climatic causality, when the erosion of the banks of the Mekong is attributed to the climate and its degradation, with no reference to the consequences of extractions and by belittling the effects of reservoir-dams (Chapter 5).

The practical conclusion is that difficulties will be overcome in favor of sustainable solutions, on the condition that the lessons drawn from defective management methods inherited from the past, and the current and foreseeable complexity, are both correctly analyzed. The development method for the delta can be found at the point where these paths cross each other:

1) adaptation of the population through a return to more traditional management measures including heavy structural measures;

2) reduction of productivity through the use of fewer pesticides and smaller harvests;

3) increased amount of space left for natural environments. This means reigning in agricultural growth to play to sustainability factors in the delta.

3.3. The Niger Delta: unlimited exploitation of black gold

3.3.1. *The deltaic zone*

The Niger Delta, the third-largest on Earth in terms of its area of 70,000 km^2, is also the delta with the largest mangrove in Africa. Its coastline, convex in shape and protected by a sandy barrier made from 20 or so bars, is sculpted by the swell and the waves; in detail, the delta has an outline that has been stiffened by fault lines; it is sharply indented by a multitude of more or less active estuaries and tidal channels. The marshy mangrove covers 6,000 km^2 of an 850 km-long deltaic “margin” between Benin and the Cameroon and subject to the forces of the sea over an area almost twice as vast. The summer monsoon rains and flooding originating from both rivers and the sea affect between 70 and 80% of the area of the wet lands. In its natural state, the delta therefore gave Africa an exceptional level of biodiversity and held global value.

The deltaic zone, made up of nine states which have a local government, has a population of more than 30 million inhabitants (in other words, approximately 25% of the total population of Nigeria); this population, the density of which exceeds 200 inhabitants/km^2, is growing at the rapid rate of 3.5% per year. Oil extracted from the delta provides 80% of the tax income for Nigeria and provides it with 98% of the revenues produced by exportation. While hydrocarbons have sustained Nigeria's growth, this comes at the price of some of the Earth's most serious social and environmental damage.

3.3.2. *The effects of the extraction of hydrocarbons on the environment*

3.3.2.1. *Almost impossible research*

The difficulty of discussing environmental impacts in the Niger Delta is due to the very significant lack of scientific information about the subject. *In situ* measurements are more or less inexistent, due to a lack of security on site and a lack of will on the part of the federal state and the local authorities, similar to large companies, to play the game of work permits. Published and often well-documented research is based on second-hand sources; one of the difficulties is that the figures produced vary greatly from one publication to another, which makes it difficult to construct an overview. In brief, there is a lack of monitoring, which explains the absence of all past and present references and explains why the evaluations are simply qualitative:

> "Monitoring is essential, but this is missing from the Niger Delta. Like monitoring the establishment of oil companies, one should make available the data concerning the terrain and its accessibility, revenue, human resources and qualified personnel. This restricts the ability and efficiency of monitoring by the government" [KAD 12].

Prior to the abrupt intrusion of oil into the region's economy in 1956, the year of the discovery made by Shell British Petroleum at Oloibiri (state of Bayelsa), 80 km to the west of Port Harcourt, the delta was a rural region that was still poor. The low level of economic interest of this region did not justify any intervention to reduce flooding, to limit erosion of the fluvial distributaries nor to improve transportation. The delta has the combined blessing and evil of being too rich in hydrocarbons, for at least approximately another 40 years if we believe the official evaluation of reserves. Oil is vital for Nigeria's development. This appears to be an economic stroke of luck for this country, which obtains the majority of its income from this, (and for the companies exploiting it, of course) even though income from oil is a danger for the states that operate under it, and a disaster for the inhabitants. There has been considerable clearing in the delta in order to install oil wells, roads and oil terminals within a primitively forested and rural landscape. The distribution of the profits from extraction and the considerable environmental damage cause brutal inter-ethnic conflicts and fighting of social groups against federal power [FAN 06b]. Six decades of exploitation have transformed it into an immense polluted industrial wilderness, in which companies have taken advantage of excessive facilitation to develop their activities:

> "Owing to government complacency and, most importantly, connivance, there has been a total disregard for the social, political, economic and environmental sensibilities of the Niger Delta peoples by petrobusinesses operating in the region including Shell, Chevron, Agip, Total, Elf and Mobil" [ODO 11].

3.3.2.2. *The scale of the watershed*

The causes of the difficulties encountered in the Niger Delta are primarily related to the developments in the watershed, independent of the factors associated with oil. The hydroelectric dams constructed on the upper Niger and the Benue hold 30 km^3 of water; they also retain 70% of the sediments transported by the river. The initial benefits of their construction in terms of flood control are partly eliminated by infilling of their reservoir. Several additional effects arise from this.

3.3.2.3. *Coastal dynamics*

If we relied on simply one of the rare quantified physical criteria, which is coastline dynamics, we could imagine that the delta is not doing too badly: the retreating sectors constitute 60% of the modified surfaces, compared to 40% which are currently being formed [ADE 10]. These figures, based on the comparison of low-resolution Landsat satellite images within a sequence from the period 1986–2003, were recently contested by a new study, this time in relation to a sequence from the period 1986–2010; on the contrary, the delta appears to be gaining ground on the ocean. The receding sectors seem to correspond to the absence of mangroves, which means that they are highly vulnerable in the context of rising sea levels [OYE 16]. These contradicting results do not appear to show any particular effects of sediment deficit, subsidence and rising sea levels. However, the situation is very serious in the opinion of A.C. Ibe, according to whom 7,000 km^2 have been lost because natural subsidence is no longer sufficiently compensated for by alluvial deposits on the surface of the deltaic plain. Erosion of riverbanks and coastlines has taken place due to the sediment deficit; coastal recession has reached several tens of meters per year. The shorelines of estuaries, lagoons and sandy formations are considered to be highly vulnerable. In the Niger Delta, more than 40% of the coastline is highly vulnerable, even more so than in previous estimates. To summarize, the assessment must be revised from top to bottom:

– reduction of the flow rates during the dry season encourages penetration of the salt wedge into the estuaries;

– geological subsidence takes place at a rate of 5 mm/year on average in the southwest part of the delta, but it has accelerated to reach 25 mm/year, or locally to 125 mm/year in response to the extraction of hydrocarbons [IBE 96].

The difficulties are also related to the rise in sea levels. If this were to reach 1 m, the number of displaced persons would be 600,000; this figure, which dates from 1995, would today be greater, taking into account the increase in population [MUS 14].

3.3.2.4. *Serious environmental deterioration*

The influence of oil infrastructures is considerable, with more than 600 concessions including 40% offshore, more than 5,000 wells and 7,000 km of pipelines. In addition to this, clearing has taken place in order to place the tomography lines, explosives have been used and substances have been released in suspension which increases the turbidity and reduces photosynthesis. Lastly, oil and gas are transported to refineries (Port Harcourt) and sea terminals, which generate their own impacts.

The most serious consequence of extraction is oil spills, caused by leaks from wells, explosions involving wells in the sea, vandalism of oil pipelines, sabotage and ageing of infrastructures, leaks from storage tanks and deliberate releases from sea tankers. Since the end of the 1950s, the losses have been estimated to be 1.5 Mt for thousands of incidents or accidents, or in other words, for the Niger Delta, 50 times the volume lost by the Exxon Valdez oil tanker in 1989 in Alaska. The figures are surprising, with a record of 570,000 barrels[5] in the Forcados estuary. These releases are carried out indifferently in the marshes, the mangrove of *A. germinans*, the rivers and the sea, causing plant species to die as well as the shrimps, crabs and fish which make up the basis of the fishing economy.

While gas pollutes the atmosphere, oil exploitation is associated with direct releases into the environment, which makes water consumption dangerous for the population. The toxicity of waters and soils is due to the interaction between the releases (metals, radionuclides and soluble organic substances) and the high salinity of the environment; combustion of waste at the surface destroys the soils. Lastly, wealth production is not accompanied by the creation of facilities such as paved roads, health services or electricity [ITE 13, NWI 06]. The living conditions for some communities have deteriorated so much that they have had to be relocated.

However, the extraction of natural gas is not to be outdone, with more than 120 gas combustion sites and direct losses in a gaseous state into the atmosphere; with a volume of 0.5 km^3/day, this is the highest release in the world as a proportion of the production. 75% of production is thus released without being used in industry.

5 A barrel contains 160 liters.

This practice should have stopped in 2008, but in 2013, the losses still reached 11.5% of production (45 billion KW of heat per day), worth two billion dollars (the equivalent of 3,500 MW of electricity, whereas 75 million Nigerians do not have access to the network). The environmental consequences are a significant contribution to the greenhouse effect (45 million tons of CO_2/day in 2015), to weakening of the ozone layer and to acid rain, which is in turn responsible for acidification of surface waters and damage to animal resources and vegetation. Lastly, we can add to this the effects of toxicity on human health (cancers, respiratory problems, etc.) and on animals, as well as excessive heat which damages plant vegetation structures and agricultural productivity. Gas combustion is therefore both an enormous economic loss and a source of pollution.

3.3.3. *Serious social and political stakes at play*

The negative effects of exploitation of hydrocarbons (oil and gas) affect populations who are deprived of their traditional resources and who do not receive investment from industrial profits rechanneled into development. Many inhabitants have lost their land and their fishing grounds to the hands of companies; financial compensation for loss of property or use is judged to be very slim. The crisis has been made worse by the fact that lands as resources are badly protected by the State, which took ownership in 1978.

Hydrocarbons have certainly benefitted Nigeria, but they are responsible for deteriorating living conditions of the delta's inhabitants, who have been sacrificed in the name of development. Successive governments are blamed for not carrying out or imposing a minimum of measures in terms of justice and development. In a more general manner, the delta region has received a decreasing portion of the financial resources that are allocated by the State, in such a way that the economy, with the exception of hydrocarbons, has declined over the course of the decades. The inhabitants consider that the federal government is in league with multinational companies, in particular in the field of pollution management. This has led to a hostile attitude towards the operating companies, with acts of vandalism and sabotage, physical opposition to prospection and exploitation, and hostage-taking to demand a ransom. The country's youth is rebelling against injustices and the poverty brought about by negligence and corruption. The expression *poverty qua poverty* is used to designate absolute poverty in Sub-Saharan Africa, in particular in this region where the average life expectancy has fallen to 40 years [IKE 09].

At great cost, the reaction of the authorities has combined incomplete aid programs, violent acts of military repression (supported by certain Western countries) and rare acts of amnesty [AWO 13, OLU 12, OVI 10]. Programs involving the United Nations Environment Program (UNEP) have been put in place on land belonging to the Ogoni and to Ijaw fishermen, land that has been heavily polluted by oil leaks and bruised by the repression of rebel movements by the federal government. In the case mentioned, it is not clear what role the international companies such as Royal Dutch Shell will play in the attempts to repair the damage, if one day remedies do indeed become possible. In 2010, UNEP officially admitted, on the basis of a report paid for by Shell and commissioned by the Nigerian government, that the company was only responsible for 10% of the damage done in 40 years (the spillage of nine million barrels, in other words twice as much as the leaks from the Deepwater Horizon platform in the Gulf of Mexico), while the rest was attributed to the communities, due to their actions of sabotage and the armed robberies carried out on the oil pipelines by "bunker gangs" (Figure 3.9). Thus excused, or at least nearly so, by the United Nations, the company was not required to make repairs [VID 10][6]. However, while the inhabitants feel that the underlying causes of the deterioration of their environment are external to the delta and consider themselves to be victims, are they exempt from all responsibility? The answer is difficult and the causes of the oil spillages "have never been the subject of independent and effective verifications or measurements"; it would be necessary to fully investigate by means of in-depth analyses into the winners and losers of the reduction in ecosystemic services [ADE 11].

Figure 3.9. *In-filled channel in the Niger delta. The mangrove has been destroyed by oil leaks. (Source: AFP/Getty Images – Pius Utomi Ekpei). For a color version of this figure, see www.iste.co.uk/bravard/sedimentary2.zip*

6 Robberies affect around 10% of production. The lack of jobs causes some of the inhabitants to turn to actions that are sometimes very violent, supported politically by unscrupulous local elected representatives and very severely repressed by the army. For a long time, Nigerian oil has been considered to be important for the security of energy supplies in the United States.

3.4. The Indus Delta, dramatically dried out

3.4.1. *The delta and its coast*

In its natural state, the Indus Delta covered an area of 17,000 km^2, of which 10,000 km^2 were subjected to the action of modest tides, with a range of 2.7 m, and to the dominant action of waves that were produced by the winds of summer and winter monsoons[7]. The sea redistributed the deposits on the coast and flooded the mangroves and muddy zones. In its natural state, the solid and liquid flow rates, which arrived along 16 large estuarian branches, allowed correct deltaic activity over an area of 6,000 km^2 and were at the origin of the rapid progradation of the delta, estimated to be an average of 44 m/year for the last two millennia, and more than 200 m/year at the mouth of the main channel in the 19th Century. Mobility was high among the changes in the relative importance of the channels and the effect of the differential mobility of the deltaic floor in response to neotectonics and earthquakes.

3.4.2. *The deleterious effects of dams on water and sediment fluxes*

The initial impact of reservoir-dams and the diversion of water to irrigated perimeters involve flow rates and the solid load. Let us recall that the annual volume of water reaching the sea was 150 km^3 in the middle of the 19th Century. Initially planned to be 45 km^3 downstream from the Kotri dam (1961), the annual liquid flow rate should be 21 km^3 following an agreement on principle between India and Pakistan that was concluded at the time of the construction of the Tarbela dam (1976); it was reduced to 12 km^3 after the 1991 Water Agreement. This enormous reduction is the primary cause of the "progressive death" of the Indus Delta. In practice, years can go by in which the minimum flow rate is not produced (it reduced to 1 km^3 in 2001 and 2 km^3 in 2002), and the water is only available during exceptional monsoon years. As for the solid flow rate, this has been reduced by 250 Mt in the middle of the 19th Century to approximately 13 Mt at the beginning of the decade 2000–2010.

The Indus now only reaches the ocean for two months each year. The large number of distributary channels has been reduced to give way to one single channel, the Khobar Creek, which now only has a noticeable effect on 10% of the original area of the delta. Tidal energy, reinforced by the rise in the level of the ocean, has become the main driving force for the reorganization of landforms; it penetrates inland to a distance of 225 km in the current channel, whereas the old branches have been reworked and widened by tidal flows, and it invades basins that have collapsed due to tectonics. The shape of the delta also reveals the contribution of waves to its construction (Figure 3.10).

7 The river within its basin was presented briefly in Volume 1, Chapter 4.

3.4.3. *A serious environmental, economic and social crisis*

Located in an arid, subtropical environment, the Indus Delta is home to 20 million people (including the coastal town of Karachi), of which approximately 100,000 are dependent on fishing. The delta has lost almost all of its fluvial input and, now tide-dominated, it has suffered considerable impacts since the Kotri dam was commissioned. These impacts, which have become worse since the Tarbela dam was commissioned, are particularly serious.

Erosion of the delta is expressed by an annual loss of 47 million tons of sediments and by a loss of surface area of 12.7 km^2/year due to the effects of the sea.

From a terrestrial point of view, the very significant reduction in monsoon flooding since the construction of the Kotri dam (1961) has eliminated overflow from branches of the Indus at the surface of the deltaic plain. Confinement of the main channel between its levees means that the deltaic zone is now only active over 1,200 km^2. A result of drying-out of the deltaic and coastal plain is the simplification of the mangrove to solely *A. marina*, a single species, stunted and fragmented; it is rapidly receding for ecological reasons and due to overexploitation by locals in search of wood. The unavoidable consequence of the minimal protection provided by the mangroves is greater exposure of the coast to storms from the Arabian Sea.

The other effects of the delta drying are the invasion of salt water during high tides, salinization of the ground and increased rarity of swampy areas. Today, salinity ranges from 38 to 42 g/L in the creeks, the equivalent of the *arroyos* or small estuarian branches in the Mekong Delta, which is stronger than the 36 g/L in the Arabian Sea. Fishermen record a 90% reduction in the production of fish, shrimps and other crustaceans in the creeks; this is due to the significant degradation of the mangrove, which no longer makes its nourishing contribution. Concerning the productivity of aquaculture activities in the marine environment, this has fallen, as has the catch per unit of effort.

On the deltaic plain, the water is too salty for the cultivation of red rice, to the extent that the latter has all but disappeared, in the same way as sugar cane. Production of milk and butter, held in high regard by the cities, has fallen. The number of goats and sheep has significantly decreased; only the vast herds of camels graze the relics of the natural vegetation. This situation is not sustainable for the population of the delta, which lacks drinking water on the surface as well as from underground sources. Consequently, since the early 1960s, the downstream part of the delta has seen a population exodus from those leaving in favor of the capital Karachi, located on its Western edge [ASI 00].

In addition, these trends strongly risk being exacerbated by current climate change, since Pakistan is made all the more vulnerable by its hot climate, where any increase in temperature will have an effect on human health (models envisage an increase of 4°C in the temperature from now to the end of the 21st Century). The meteorological stations in the country already record a significant increase in winter temperatures, an increase in evapotranspiration and heat stress; the shorter time period for agricultural production in winter is worrying, because seeds and bananas will no longer have time to ripen. The country fears the strengthening of cyclonic activities and an increased variability of monsoon precipitation [RAS 12].

Figure 3.10. *The Indus Delta at the end of the dry season. From east to west: the Gulf of Kambhat with very large tidal ranges, the Gulf of Kutch, the salty areas of the Great Rann of Kutch and lastly the lower irrigated Indus Valley. The natural vegetation has greatly deteriorated. (Source: Chelys, eosnap.com). For a color version of this figure, see www.iste.co.uk/bravard/sedimentary2.zip*

In summary:

> "The Anthropogenic Indus Delta is hardly a true delta any more, it receives too little water and sediment from the fluvial system, and tidal processes have taken control of the environment. In effect, it is a relict landform from the pre-Anthropocene time" [SYV 14].

It seems impossible to hope for a return, even partial, to the natural state. However, Pakistani specialists recommend no more reservoir-dam constructions upstream of the Kotri dam in order to avoid further reduction of the flow rate reaching the delta. They also recommend releases of freshwater to raise the instream flow to the level dictated by the 1991 Water Agreement; their objective is to reduce the salinity of waters and soils, and to replant species that are likely to protect populations against cyclones and tsunamis. The World Bank, for its part, would be content with half of the flow rate agreed on in 1991 [CHA 11].

3.5. The Ayeyarwady[8], initial symptoms of imbalance?

3.5.1. *Burma, a country on the cusp of development*

If there is a delta for which behavior close to natural processes may be hoped for, it is the Ayeyarwady, a little-known river, at least in the eyes of the wider public. This was realized back in 2016, on the cusp of commitment to a study of the geomorphology of the river for the World Bank. At the dawn of the hoped-for development of Burma, a potential source of impacts, and in the face of hydroelectric development instigated by China on the upper part of the river, it was necessary to draw up an assessment of the watershed's operation, of the river and of its delta, in the same way that it was necessary to highlight the constraints that the developers would be faced with and to strengthen the ability of governmental agencies to control projects.

In 2007, the Myanmar government had signed an agreement with China for the construction of seven large dams with a total power of 13,360 kW, including 3,600 just for the Myitsone dam, ranked 20th in the world for its height of 150 m and with a reservoir of considerable size. The location of these structures in a highly seismic zone, on a very active fault line, was worrying for Burmese decision makers, who decided to stay the project, while the ethnic minorities in the mountains and the NGOs were in turn worried about the possible impacts of the chain of dams on biodiversity, considered to be remarkable, in this Indo-Burmese region, on migration of migratory fish and on the retention of sediments. Moreover, intense exploitation of forests, jade mines and fluvial sediments and the uncontrolled rise of industrial agriculture should be taken into account.

8 The Ayeyarwady is the name used in Burma for the Irrawaddy.

3.5.2. *The Ayeyarwady, an enormous conveyor belt*

The Ayeyarwady is a river that is 2,260 km long if the N'Mai Hka is taken into account, which is the longest of its two branches in mountain regions which join up at a distance of 1,740 km from the Indian Ocean; it receives large tributaries, of which the most important is the Chindwin, which drains the western part of the basin and the area of which approaches 415,000 km^2. The Indo-Burmese range and the Himalayas create a strong orographic effect on the volume of monsoon rains, where the pluviometric regime, subject to the summer monsoon, conditions the hydrological regime of the river, with its high waters in July and August and its low waters in February and March. The annual flow rate of the river at its mouth has recently been estimated at 380 km^3, a value that is lower than the 420 km^3 measured by the English in the 1870s – but are these evaluations reliable? In terms of sedimentary transport, since the 1940s, it is recognized that the upper fluvial basin provides little, contrary to the large tributary that is the Chindwin and a region of the southern channel, the Central Dry Belt, whose dry climate favors erosion of the slopes [GOR 1885, STA 40]. The values recalculated in the decade 2000–2010 make the Ayeyarwady the seventh largest supplier of sediments in the world to the ocean with 365 Mt/year, a value affected, however, by a large margin of error [FUR 09, ROB 07].

3.5.3. *The delta: crisis or stability?*

Construction of the Ayeyarwady Delta began at the end of the last glacial period, in a tectonically stable estuary dug into a fan shape; the deltaic plain was constructed from 6,300 years BP by the input of mud deposited between the fluvial channels of sandy deposits, which exhibit historical avulsions; for four millennia, the coast has featured sandy ridges that mark the progression of the delta towards the sea. Satellite images reveal an average progradation of 3.4 m/year in the last century and a high level of turbidity in the sea on the continental platform, the site of large amounts of deposit. This is the typical case of deltas that are subject to the actions of strong tides (of an amplitude of 2–7 m) and also to sea currents exhibiting a west–east drift. The influence of the tide is felt up to a point upstream of the delta, where its branches separate, 290 km from the Andaman Sea. The tide penetrates via numerous estuaries, whereas the action of waves produced by the monsoon from the southeast is of lesser importance. Progradation of the delta into the sea has, however, reinforced the redistributory action of waves of heights ranging between 1 and 2 m. The delta, which covers an area of 35,000 km^2, prograded on average by 7.8 m/year between 1974 and 2015, but local values vary (Figure 3.11). The cyclone Naris (2008) caused a generalized recession of the coastline, which leads to fears of high levels of vulnerability, increased by mangrove clearing in favor of rice paddies and aquaculture, in the same way as in the Mekong Delta [BES 17].

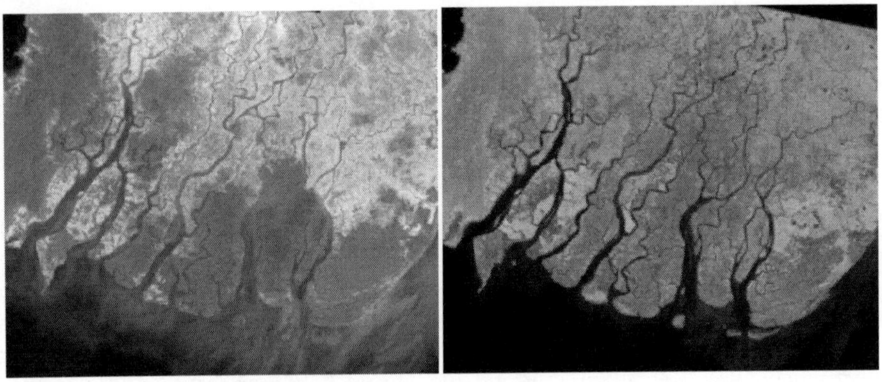

a) b)

Figure 3.11. *The Ayeyarwady Delta in the dry season, with its subparallel distributaries. It lost 1,685 km^2 in surface area of mangroves between 1978 and 2011 due to expansion of rice cultivation and aquaculture. (a) Landsat 1, January 1974. (b) Landsat 8, February 6, 2017. (Source: USGS, earthshots.usgs.gov). For a color version of this figure, see www.iste.co.uk/bravard/sedimentary2.zip*

What diagnosis can be given for the delta's health? According to a number of authors, the Ayeyarwady Delta receives fluvial and marine sediments that aggrade its surface, but it does not receive enough of them. The deficit in sedimentary inputs (presumed, at this stage of knowledge) is thought to be due to the dams that have already been constructed in its basin, but the sedimentary congestion of its bed upstream of the delta causes doubts to be cast on this. It is probable that the fluvial dykes that have been constructed over the last 150 years have polderized vast areas dedicated to rice cultivation; these are therefore relatively well protected from sedimentary input [ANT 17]. In fact, the observation of a reduction in delta aggradation caused by the river, and of subsidence accentuated by the effects of dams in a context of rising sea levels [BES 17], is not given, because there is still very little, if any, documented evidence of these factors; modeling of the effects of dams constructed on tributaries certainly provides a reduction in sedimentary inputs, but these figures should still be treated with caution. In addition, opinions vary concerning the observed trend in the longer term, between relative stability and recent recession of the coast [HED 10]. Predictions for the reduction of inputs should in principle cause recession of the coast, but a recent study declares that, "unlike most other deltas across the world, the Ayeyarwady has not yet been affected by dam construction, providing a unique view on largely natural deltaic processes benefitting from abundant sediment loads affected by tectonics and monsoon hydroclimate" [GIO 17]. The key to the atypical operation of this delta, which progresses little, or is even receding according to certain authors, is not a response to the construction of artificial reservoirs, but a transfer of sediments off the coast in a context of high levels of tidal energy and, to a lesser extent, of waves [GIO 14]. We can add that dyking has certainly helped sand to descend towards the sea and has aided the expulsion of a significant proportion of mud towards the deltaic plain and even off the coast. This scenario is in phase with the observation that the river load is considerable and is increasing.

3.6. Conclusion

The large deltas in Southeast Asia, which we have selected as examples, present different situations that offer instructive similarities and differences.

To summarize the similarities, the deltas in question are of course those of which the hydrology is subject to the rate of flood events produced by the summer monsoon and of which the sedimentary load is related to the erosion of young, productive mountain ranges. Nuances do of course exist between the systems presented, but the rivers can in some way be described as cousins. Then, the basins, including the Ayeyarwady basin, are drained by international rivers, which induce a possible domination effect of upstream countries over downstream countries that are established on highly populated deltas. This is indeed the case. Chinese water management carries more and more weight for the Brahmaputra, the Mekong and even the Ayeyarwady, whereas India greatly affects the Ganges while reinforcing their development projects on the Brahmaputra. Concerning the Mekong, Thailand, for reasons known to itself, and Laos, which falls within the economic realm of China, have abandoned the concertation policy that was implemented in 1995 in favor of an "each to their own" idea, which carries heavy consequences for the delta and therefore for the downstream countries of Cambodia and Vietnam.

In a way, the constraint that applies to these deltas tends to present increasing and very worrying similarities. The countries in the fluvial basins concerned want to have their legitimate part of the development and rely on the rise in hydroelectricity production. This type of energy was strongly critiqued at the end of the decade 2000–2010, to the point of losing funding from the World Bank for a while, but international awareness of climate change rekindled interest in carbon-free energies, particularly hydraulic energy, which returned to fashion, its reputation repaired by ecological virtues that had previously not been recognized. Consequently, this resulted in a rush for hydraulic energy on reservoir-dam sites, with the undesirable effects of breaking up fluvial linearity with respect to movements of migratory species and alarming forecasts in terms of retaining sediments, at least in the case of the Mekong.

The fact that Cambodia and Vietnam have widely benefitted from extractions, which they themselves have organized or tolerated, weakens their position, which is otherwise legitimate. However strong or weak the diplomatic and military power can be in Vietnam, the country that is the most affected by the degradation of the delta, it is certain that China puts so much pressure on resources in the East China Sea that Vietnam cannot fight on both fronts. Bound hand and foot, this country will have to manage the effects of dams in the basin, in addition to those of climate change in the sea. Fortunately, it can rely on economic growth and on technical and material help from numerous countries (Europe and Australia). On the other hand, it is unfortunate that one of the poorest countries in the world, Bangladesh, is put under severe pressure by countries upstream and that this pressure features very little in the media. In any case, this poor and overpopulated country is incapable of producing the resources that would allow it to protect a growing and exposed population from the effects of climate change, such as the worsening of cyclones and flooding originating from the sea.

4

The Aging Delta of a Country in the New World, the Mississippi

The Mississippi Delta, encompassing an area of 25,000 km^2, entirely belongs to the state of Louisiana, as much from a physical standpoint as a cultural and human one. An integral part of its identity is the historical city of New Orleans, constructed in a very dangerous and debatable location, even though its location appears *a posteriori* to be founded on economic logic. The paradox (one of them at least) is that the error of that original decision has been followed by other economic and technical choices considered to be just as harmful, the fact that the weaknesses that initially applied to the delta's capital city have become those of the delta as a whole. In short, the water is rising. Of course, this is because the sea level is rising, as we have known for two centuries, but above all because the delta is sinking. And it is sinking because decisions made at the scale of the Mississippi basin have caused the inputs that are essential for the life of the delta and because other decisions, made at the scale of the delta, did not foresee that the latter, incapable of regenerating itself, would sink.

4.1. New Orleans: an "inevitable city on an impossible site"

4.1.1. *"Discovering" the river*

The use of the word discovery is in itself a mistaken approach, because it neglects the fact that Native Americans have lived on its banks for thousands of years. The Mississippi was the Father of Waters or the Great River for the Ojibwa nation; it became Big Muddy for the inhabitants of New Orleans [CAM 08, MAR 08, POW 13, SAX 27]. The Mississippi is a river in the New World which, curiously, was first explored by Europeans upstream before being explored from downstream, starting from the Gulf of Mexico. The mouths of the Mississippi featured on a map, drawn a little after Christopher Columbus's fourth journey (1502), and a long time before the disastrous expedition by Hernando de Soto (1541–1542) who explored the lower valley near Memphis via an inland route, which shows that Columbus was indeed the true European "discoverer" of the mouth of the Mississippi. However, putting a hunch to a map is not enough: nearly 200 years were necessary for the river mouth to be properly explored by Europeans by sea in the midst of interlacing muddy outlets. The French explored one mouth during Radisson's expedition in 1655, starting from Quebec; Joliette then made an attempt in 1673, before the success of René-Robert Cavelier, Sieur de La Salle, who descended in a canoe from upstream all the way down to the mouth of the river in 1682. Following this, he staked a claim to Louisiana, naming the future US state after the French King Louis XIV, but he did so from the river, not from the Gulf. De La Salle failed when starting from the Gulf in 1684 and the credit for the European "rediscovery" of the mouth of the Mississippi in 1699[1] should go to Pierre Le Moyne d'Iberville, especially for having journeyed for the first time upstream to the Native American site of Baton Rouge[2]. The unpredictable nature of this coast, with its barriers and sandy islands, its bays and multitude of channels, partly obstructed by woodlands, helps us to understand what a natural delta, loaded with mud and tree trunks, it really was at a time when navigational instruments were still insufficient to allow route finding without mistakes.

4.1.2. *At the origins of New Orleans*

The first havens of French Louisiana were provisionally established on the coast of the Gulf by d'Iberville, prior to the current city being established on a bank of the Mississippi. This trade post was founded by the *Compagnie d'Occident* (1717), an initiative of John Law, who wanted to liberate the *Ferme du tabac* (a tax collection franchise) from its obligated submission to English tobacco coming from Virginia, with the aim of creating a French tobacco kingdom on the banks of the Mississippi, based on concessions. It was necessary for the territory to be supported by a small fortified town. The New Orleans site (thus denoted in homage to the regent, Philippe II, Duke of Orléans) was chosen by the commanding general of Bienville, the brother of Pierre d'Iberville, in defiance of official opinions from France[3]. They had instead recommended that a more judicious choice of site on which to establish the "first city" would be the Baton Rouge region, for the comprehensible reason that this was the site of the first terrace to be shielded from flooding and because the land was better suited to tobacco growing than the marshes. Its rather late date of creation, in 1718, meant that New Orleans was the youngest of the colonial cities in

1 Thanks to a map stolen in 1697 from the captain of a Spanish ship!

2 Possession of Louisiana territories slackened off in 1764, when Louis XV secretly handed over the territories located to the west of the river to the King of Spain, Charles III; it was then Napoleon who definitively put an end to the presence of the French on the North American continent by handing over Louisiana to the United States (1800–1803).

3 Powell explains Bienville's strange choice by the fact that he had obtained a land concession on the site that he imposed.

the United States, initially inhabited by a few nobles and entrepreneurs, craftsmen from mainland France, military volunteers, slaves and convicts who were forced to live in terrible conditions. Its site, at the very heart of a vast amphibious delta, was extremely unsafe, and having been chosen to cater for personal ambitions, it turned out to be the worst possible choice once extraction of hydrocarbons began 250 years later. Its site was badly chosen, but not its location, which, as the rest of its history has shown, was excellent; a city in a bad location in terms of its buildings, but in an excellent trading location, in which this latter compensated for all its disadvantages, as the German geographer Friedrich Ratzel commented in 1876 [CAM 08, KEL 03, LEW 76, LEW 18].

The banks of the rivers are higher than the surrounding land in the Mississippi Delta; they owe this to the presence of alluvial bulges formed from the silty deposits of floods that inundated them – these are the "levees", a term that is without doubt derived from the *levées* which are built along the banks of the river Loire[4]. All authors who have written about New Orleans have highlighted the constrained nature of this site at the heart of the delta, located at a distance of 160 km from the mouth. The choice of site by Bienville did have some logic: the Spanish and French neighborhoods (the *Carré*) are located on the river's levee, 4 to 5 m higher than the original sea level; the levee overlooks a stream pool, in the concave bank of a meander; the back of the levee has a shallow slope leading to the marshes with their giant cypress trees and, beyond this, to Lake Pontchartrain. The concavity of the primitive site and the map with curved, radiating streets have earned the port city the nickname *Crescent City*. Certainly, this is a perfect location for a port, but displays a total lack of knowledge of the constraints and the risk being taken: the difficulty of ascending the river and the alluvial mouth obstruction, the lack of evacuation of urban effluents, epidemics and flooding. A public area by definition, the water front is a pertinent indicator, since it allows us an insight into the complex social history and the place of the river in the construction of the city [KEL 03].

4.1.3. *An area with serious issues at stake*

The *batture* of the Sainte-Marie neighborhood is a muddy area, uncovered at low tide, which is located on the convex bank that begins upstream of the French Quarter and of the concave bank of the original site. Different from the geometrically designed, planned quarters that characterize New Orleans, the batture is the public area that became the site of all trade and urban ambitions. Over the years, it accompanied the slow construction of the meander's neck. Considered to have a

4 The first levee in New Orleans, approximately 3 feet high and 18 feet wide, was the site of a road. It could have been built on the same model as the levees on the Loire in France. The construction works were carried out by local residents.

servitude that authorizes access and public activities under Roman Law, adhered to by the Spanish and the French (Creole peoples), the batture is an entrance to the city, a market, a port, a rubbish tip and a promenade, all at the same time; a place where fluvial deposits have constructed the city's substrata. This European view of things, which has become customary, and was supported by then-president Thomas Jefferson in Washington, contradicts the desires for privatization of the batture expressed at the beginning of the 19th Century by the local owners, directed by Livingston, the entrepreneur and Secretary of State; for him, in the same way as for "new Americans", private property is a sacred principle of the United States Constitution, on the grounds that a pointless, useless and dangerous land must absolutely be put to good use (1808). In the war over the batture, alongside most of his fellow citizens, the city's representative in Congress maintained that the batture would be a great port, the only one possible, and a place of great aesthetic value; a modern *sense of place* was thus precociously created in New Orleans.

Set up in 1811, the Fulton group brought together investors who sought to promote steamships on the river. They exercised a monopoly by maintaining the batture as a public location in order to make it their port (which in fact came down to requesting concession of the coveted terrain). This obvious and rather hypocritical contradiction was very much present in approaches at the time, with investors generally seeking to maximize the benefits drawn from public resources while trying to improve nature using unnatural techniques. The technicist's utopia necessarily needed to serve the public interest. In principle, monopolies did not receive the relative scorn that they experience today; in the mentality of the time, they were considered to be factors for development and environmental improvement. Since 1817, however, the Fulton group had to concede to a competing entrepreneur, Henri Shreve, and to public opinion in the valley, which was clearly in favor of totally free trade on the river. At the beginning of the 1860s, more than 250 steamships guaranteed movement around the third largest port in the world. Growth of the city was such that it appeared to have demonstrated its capacity to carve civilization out of the wilderness, to manage nature entirely. Certainly so, but to the detriment of removing the real constraints of the Mississippi's surrounding environment. In the middle of the steamship century, New Orleans had become "the quintessential river city" (Shreve) and "the technology appeared finally to have delivered the Mississippi's promise" in a city which was becoming highly self-conscious [KEL 03]. Around 1820, the Mississippi became fully incorporated into the United States. The federal government had acted on the increased demand and decided to finance development of the river as required.

Very quickly, a series of constraints came to light in the fluvial city, while full awareness of the changes did not dawn even on the city itself, with the exception of the destruction of neighboring forest upstream from the city, brought about by the energy requirements of steam-powered machines. In a first instance, by placing the riverbank and the city a certain distance apart, the fluvial dynamics had progressively pushed traders and American owners to make a request in 1835 for the development of a deep-water quay, which would be free of tidal constraints on the docking berths; a project contested by the Creole peoples who were worried by the visibly growing power of entrepreneurs a short distance away from the historical center. The victory of trade and order over the traditionally vague concept of a public space created a new stage in domestication of the riverbanks and division of the city into districts based on ethnicity, where the Americans had somewhat broken away from the historical districts. The city councilors had just put in place the rigidity that suited the rise in trade and weakened the city's capacity for adaptation, making it "less flexible to a changing environment". Certainly, the city councilors put basic infrastructures in place in order to move the waters away, and had succeeded in their management of the levee since the 1790s, but:

> "New Orleans' quasi-liquid landscape continually mocked European efforts to erase nature from the picture" [POW 13].

Also, north–south trade, which had focused on the river for decades, suffered competition from canals and railways that had rapidly drawn up a strange, complex web leading to the river. Then, the limits of technology became clear. The race for speed, permitted by the conversion to high pressure steam in the 1840s, turned out to be very dangerous for ship and passenger safety. The city, having experienced brief epidemics of yellow fever in its history (3,000 deaths in 1847), was hit even harder in 1853, with nearly 10,000 deaths. The underlying cause was a pathogen that had been brought from the Caribbean on a boat, combined with a carrier, the mosquito *Aedes aegypti*, which proliferated in the marshes and in an urban built environment that would not meet our modern criteria (urban areas were badly drained and rich in suitable habitats for mosquitoes). While unaware of the true cause of the problem, the city was only saved by the arrival of the winter freeze.

Obstruction of the river and the crisis arising from the Civil War caused fluvial traffic to collapse, to the benefit of railway companies that invested in the waterfront with their rails and their large warehouses, and cut the city away from its river. The desire to restore traffic on the river, at least in the direction of the Gulf, increased in strength during the Reconstruction phase. In 1869, technological assistance was requested in order to avoid infilling the downstream channel at the Head of the Passes where it divides into several branches in the form of a duck's foot. While the strength of traders (in terms of the New Orleanians) backed the idea of a canal to double-up the river all the way to the coast, Congress and the US Army Corps of Engineers (USACE)[5] finally accepted the project by civil engineer James Eads, to

5 The US Army Corps of Engineers is a federal agency with its school at West Point. Created in 1802, USACE was authorized in 1824 to construct bridges and canals and to intervene on the Ohio and the Mississippi Rivers. It was deeply involved in the Civil War, both in military

adapt the channel by constructing jetties designed to narrow it and increase the energy of the flow rate. From 1875–1879, Eads' construction work allowed the Southern route to be opened to trade and reinstated navigation heading down the river, carrying cotton and cereals, the symbol of the new agricultural economy. During the years 1880–1900, the city had not lost hope of recovering its waterfront. Citizens made a request, especially modernist in nature, for this to happen. They had the intention of restoring control of public power, taken back from the railway companies. The Dock Board decided to construct a landscape made of quays, cranes and gigantic warehouses that isolated the city from its river, a sign that on the contrary to New York and Chicago, the economic forces at work in New Orleans, obsessed with development and doubtlessly marked by very difficult crises, denied any value, even symbolic, to the fluvial nature of the Mississippi waterfront. The river was very much an economic tool and other uses were excluded. It is true that today all the port infrastructures in Louisiana (in particular, the fluvial complex) guarantee the annual circulation of 450 million tons, i.e. practically 20% of sea trade in the United States.

4.2. Floods and protection of the lower Mississippi valley and the delta since 1717

4.2.1. *Initial protections*

The lower Mississippi plain, several kilometers wide, is bordered by a double row of modest cliffs, named bluffs, ranging from 15–60 m in height, and providing a natural limit to the spread of flooding. Before the construction of artificial levees, these poured freely into the floodplain, in such a way that water and suspended sediments circulated in parallel with the river or returned to it when the waters receded; part of the sediments aggraded the lateral basins and the displaced waters joined the river further down. The Mississippi was bordered by natural levees, but the lateral mobility of the river destroyed them as their construction continued, in such a way that they remained modest in size; they were reinforced by artificial structures that were themselves qualified as "levees". Cartography of the Mississippi is remarkable in more ways than one. The geologist Harold Fisk's interpretative cartography is the most surprising, with its systematic representation of the Holocene course of the river in the Mississippi meander belt, between Cape Girardeau (MS) and Donaldsonville (LA) [FIS 44] (also refer to Volume 1, Chapter 5).

Construction of the artificial levee began in 1717 in New Orleans, upstream and downstream of the city that was the center of initial organized protection. The first

engineering and in the construction of structures for civilian use, including in the Confederate branch. USACE very naturally began to focus on the Mississippi, the great military road of the conflict.

flood, in 1720, was followed by many others, on average two per decade. River engineers were persuaded that the levees had a positive effect on the depth of the river and, based on that idea, also on navigation – even though the floods caused the banks to be undercut and cave in. The height and trapezoidal cross-sectional area of the levees was for the first time calibrated all along the river by the Swamp Acts[6] in 1849 and 1850, but it could not yet be classed as effective protection. The 1849 flood had already allowed the engineer Charles Ellet to state that the new levees restricted the cross-section of the in-flood channel and increased risk of their rupture (which was an entirely new observation at the time). He recommended reinforcement of the levees and the installation of spillways towards the plain and the marshes, but his projects were too early for his time.

4.2.2. *The beginning of generalized protections*

In 1879, the Mississippi River Commission was created at the initiative of the federal government and organization of the fight against flooding saw full development. The main task was to organize and coordinate private dyking that was authorized by laws established to apply at the State scale, always with a view to obtaining positive effects on the ordinary channel. Containment of small- and medium-sized floods ensured flushing of the sandy bed, knowing that greater protection, necessarily costly, would not improve the channel. Protection of the floodplain was therefore not the priority at that time, on the contrary to the promotion of navigation which remained relevant. However, after the creation of the Mississippi River Commission in 1879, the federal government, with an effective tool in hand, implemented measures that were specifically devoted to protection (meaning independent from the measures that were destined to improve the navigation conditions, while remaining compatible with them). The devastating 1882 flood, responsible for 284 crevasses opened over a total length of 90 km to the expenditure of natural levees as much as artificial levees, played an important role in creating a certain amount of awareness, more than 30 years after the Ellet report. However, the protection policy remained the same until 1928, just after the river's largest flood where the water rose further and further between higher and higher dykes. In order to maintain only those in New Orleans, the 1717 levee was aggraded and widened in 1888 and 1914, until a barrier was created between the river and the city, a barrier which was again increased in height in 1928 and again in 1973. The success of constructions built until that time had the logical conclusion that the flooding zone along the lower Mississippi reduced over the decades. While it was 89,600 km^2 during the 1882 flood, it was then only 59,600 km^2 in 1927 during the largest (known) flood in the river's history; breaches had been created, but in a

6 The Swamp Acts negotiated spillways in exchange for reclamation of the marshes. The spillways were not opened, and the marshes were drained.

less serious manner than previously. The flood spread in particular due to openings that were kept open and which led to flood basins that were still being supplied by the river at the confluences, in places where tributaries joined them, with an upstream backwash (the principle of flooding due to backwash applied to an area of about 13,000 km^2).

4.2.3. *The 1927 flood in the Mississippi valley*

The 1927 flood affected New Orleans, which had indeed been built on a sturdy natural levee, itself raised up on an artificial levee 90 cm in height, but which was affected by upstream developments [RMS 07]; the city was protected by another fragile artificial levee near Lake Pontchartrain. The fluvial flood began on April 15, 1927, on Good Friday, and rose until it threatened the urban area near the river and the lake. The situation was all the more serious because the circuit for the electric pumping system had broken, heavy rain was falling and the underground water level was rising in the saturated basin, closed in on all sides. The city councilors chose to sacrifice the neighboring countryside for urban security by opening a breach in the right bank levee at Caernarvon on April 29, downstream from the city. The operation was carried out "in a carnival atmosphere", despite opposition from the 10,000 or so inhabitants of the affected rural areas and their armed trappers, who lost their few possessions and knew that their cattle, their equipment and the harvests, as well as income from muskrat furs and oysters, would be lost for years to come; this was indeed exactly what happened. And the New Orleans Citizens Flood Relief Committee did everything they could to block the compensation that had been promised [KEL 03].

The outcome was catastrophic for the Mississippi valley: 250 people died, 160,000 families were displaced and 65,000 km^2 were flooded. Louisiana mourned the loss of 600 houses and more than 4,000 km^2 of harvests. The authorities at various levels passed the disaster off as an act of God, or at the very least as a natural phenomenon, a claim designed to exempt them from any responsibility. Criticism was so intense with regard to the negative effects of the policy exclusively focused on levees that President Coolidge requested that the Mississippi River Commission (MRC) revise their plans for a new policy, turning back, as it happens, to the plans advocated by Charles Ellet in 1850.

> "The flood of 1927 was necessary to show everybody what some of us have known for a long time, that the valley is the victim of the most ruthless, obstinate, and monumental blundering engineering policy ever known in America... Man wants to take the river's natural storage reservoir and make no compensation for it. The river contends

it is against Natural Law and cannot be done. The river is right" – a lawyer for the spillways quoted by [KEL 03, p. 189].

4.2.3.1. *The post-flood situation and protection plans*

Other voices of authority were making themselves heard in the meetings organized by Congress between 1927–1928. The most severe came from the former head of the US Forest Service, the politician Gifford Pinchot, who was against the large-scale clear-felling practices of large companies; he condemned, as did some of his colleagues, deforestation of the Mississippi corridor and the narrow mindset of hydraulic engineers in the US Corps and the Mississippi River Commission. Like the Dust Bowl, "the 1927 flood… served a catalyst for the rise of an ecological worldview in some sections of the nation's scientific and engineering communities" [KEL 03, pp. 190–191]. These innovative designs tend to consider the river, its affluents and the major floodplain as "an integrated system, as a whole, and not as discrete parts that could be manipulated by technology" [KEL 03, pp. 190–191]. In the eyes of the historian Ari Kelman, the official line on nature's place in management of the Mississippi remained, however, a rhetorical question, where "total control" of the river remained the ultimate objective of official engineering.

4.2.4. *The Jadwin plan (1928)*

Continuation of the previous policy is demonstrated in the content of the Jadwin plan; made official by the Flood Control (or Jones-Reid) Act (voted in Congress on May 15, 1928). It was the foundation for the current plan of action. Water retention reservoirs and reforestation were not chosen because they were too costly, but the adopted principle (which was more spatially restrained than was desirable) was systematic dyking of the river, on the south bank of the lower Arkansas, on the lower Red River and on both banks of the Atchafalaya[7]. Very quickly, the large 1929 flood allowed normal function of the system to be verified; in fact, no levee section was overcome, which was indeed the objective required. In 1931, levees extended to practically 3,000 km downstream of Cape Girardeau, a town in Missouri located a few kilometers upstream from the confluence of the Ohio and 185 km downstream from Saint-Louis.

7 The levees, constructed partly from sand, were not supposed to be exceeded by the highest water levels, nor to collapse, nor to allow infiltration or saturation by water (phreatic rising of water behind the levees, known as sand boils, was feared greatly). The height of the levees, which was eight feet above the ground around 1850, was generally increased by up to 40–65 feet over the course of the 1900–1930 floods. These enormous construction works required dyke technology to be perfected, as well as innovation in the design and construction of powerful machinery that included dredgers, bulldozers and conveyor belts.

However, there were signs of evolution as, at the same time, the expansion of the flood within the basins at the scale of the floodplain was organized, with several floodways (flood corridors) arranged between the confluence of the Ohio and the river mouths, and two spillways ensured that the flood expansion basins of the Bœuf and the Atchafalaya were filled. The Bonnet Carré spillway, at the very site of the Caernarvon breach, was expected to reduce pressure on New Orleans and guarantee its protection by diverting the water towards Lake Pontchartrain, located less than 10 km from the river and 6 m lower. The floodgate, finished in 1936, had an outflow capacity of up to 7,000 m^3/s, seen through the breach when it was voluntarily opened in 1927; its efficiency was confirmed during the 1937 flood. The remarkable ability of the US Corps to dominate the river can, very readily of course, be insisted upon, but reservations remained: "humans ended the 1937 flood, not the Mississippi" [KEL 03, p.194]; the optimistic claims of the US Corps and the MRC were simply "stupidity", according to many historians, and the spillway treated the symptoms of the flood, not its causes.

4.2.5. *Current protection elements*

The current Project Flood was established in 1956 by the Mississippi River Commission on the basis of calculations by the Weather Bureau concerning the three flood events of 1937–1938 and 1950, in order to model the worst possible scenario (86,000 m^3/s at River Landing upstream from the delta's floodways). This plan, which protects 4.5 million people, was designed on the basis of dykes, fixed and fuse-plug levees, and dredging. In 1944, the complete system of reservoirs in the upper basin was added, in compliance with the 1944 Flood Control Act; it was likely to have a positive effect downstream [MRC 07]. The successive plans established by the USACE have been severely criticized, because they had not planned enough spillways and accelerated the flood wave by cutting off meanders. In fact, construction works shortened the channel by more than 240 km, which reduced the level of the flood by 4.50 m in the Greenville sector but made it worse downstream. Lateral erosion has, however, partly restored the initial length of the river [BAR 97].

4.3. The "deltas" in the lower Mississippi valley, from wilderness to the current crisis

Geography allows us to define a delta in simple terms: this is the zone in which a river is divided into branches and constructs a triangular area of land with its alluvia, which has been reclaimed from the sea. The notion of a delta is more complex in the lower Mississippi valley. From the Mississippi Delta and the Arkansas Delta, located upstream, to the Mississippi River Delta, which is the delta itself, the Mississippi has a somewhat disrupted geography in terms of its toponymy. On what criteria are these distinctions between deltas based? It would be helpful to look in more detail at the two truly fluvial zones, which are qualified as "deltas" in common language, although they have only some of the characteristics of a true delta, and certainly not the characteristic mouths. These are in fact amphibious zones, whose economies, built up over centuries of Native American occupation before colonists arrived and exploited it, have numerous factors in common.

Two publications give us useful information about the Arkansas Delta and its associated twin on the left bank [COB 94, GAT 93][8]. Very similar to the Mississippi Delta, which is its counterpart to the east of the river, the Arkansas Delta, "the deepest of the Deep South" [GAT 93], is the amphibious region that occupies the alluvial Mississippi plain in the eastern part of the state of Arkansas. This area is considered a delta in local toponymy, because it has stereotypical features of one, but it is not really a "true delta". It is a landscape, cut in two along the axis of the valley down Crowley's Ridge, in which the river, its former Holocene channels and the lower course of the Arkansas River have built up very rich alluvial materials and soils on their low-lying lands (their Lowlands), while causing instability and disasters. Thanks to the wet climate (1,300 mm) and the warm summers and the winters that do not dip below freezing point, the "Pre-Columbian" valley bottom was covered in a very rich alluvial forest of hardwoods (oaks, hickories, sweet chestnut trees, elms, maples and gum trees) that has become well-adapted to the long duration of the period of submersion which falls into the category of Southern Floodplain Forests, with the exception of the reworked fluvial fringes and the disturbance to ecological succession. This forest slowed down the floodwaters, on the contrary to the lands that were cleared later on. This part of the country is a mosaic of channels, bayous*, oxbow lakes*, ridges and swales*, where the latter are occupied by marshes and stagnant environments suitable for the decantation of floodwaters (slackwaters with black clayey soils). It was, at the time of the Native Americans and the first colonists, a hunting ground of birds and waterfowl, with waters teeming with fish.

The Mississippi Delta and the Arkansas Delta were the sites of the primary windfalls (the first "bargains"), which seemed inexhaustible in the eyes of the colonists due to their abundance, and required little efforts on their part to obtain satisfactory income, for example thanks to the trade in skins. However, they overlooked the role that had been played out for centuries and before the 1730s by Native Americans from the Natchez, Choctaw, Chickasaw and Biloxi tribes (to mention only a few), who sought out the sandy-silty soils on the levees that had been fertilized by floods and were easy to work, for the cultivation of corn and beans.

8 The Mississippi Delta extends from Memphis to Vicksburg over the entire fluvial landscape of the state of Mississippi (latitude 35°N to 32°N). The Arkansas Delta has the equivalent extension in the state of the same name (latitude 35°N to 32.4°N). This fluvial landscape extends along both banks of the Mississippi in Louisiana.

This is to say that the forest had already been pushed back significantly by fire and was no longer a primary in the 17th Century, even though it retained a magnificent diversity of flora and fauna. Agriculture had opened vast clearings and the space dedicated to prairies had been modified. Colonists even observed bison in the delta's open areas that they were exploring [SAI 05].

The secondary windfalls came from clear-felling the primary alluvial forest, with a peak between 1850–1860, then from the single-crop farming of cotton. Wood industries were centered around Vicksburg (bald cypress was in high demand) and enjoyed a boom in the decades 1880–1900 to satisfy demand from the north of the United States; the wet Memphis forest became the "worldwide hardwood capital", while within the delta, the forest became the target of the "fever" for Spanish moss, an epiphyte with multiple uses. The railway extended its networks into the heart of the delta and, taking over from floating, led to wealth being transported to the North, because the lower valley and the delta had become the last region containing hardwoods. Against the constraints of cotton fever and the demand for pulp to make paper, regeneration of the clear-felled forest had become impossible, even within the delta. The forest was considered to be simply an "obstruction to agricultural lands" and was required simply to be eradicated to make space for the softwood lumber industry. At the beginning of the 1930s, 60% of the delta's forests had been converted into an agricultural landscape; the hardwood forest only subsisted in the wettest and most far-flung areas and on the batture located between the river and the levees, to the extent that they now only represent 5% of the delta's surface area. The landscape described by William Faulkner in the story "Delta Autumn" in *Go Down, Moses* (1942) was already no longer the wilderness of times past; this had disappeared, whereas land drainage instigated by USACE continued in order to benefit mechanized agriculture and catfish farms.

In the 1830s, treaties had deprived Native American nations of their lands and given them to the federal government, which used them to open a new agricultural frontier and to create rural townships. It was at this point that the idea of the "paradise on earth" of the wet wilderness of the Arkansas Delta disappeared. This area was, at the beginning of the 19th Century, already a land of rich cotton plantations (the "Cotton Belt" deserved its name from 1815 onwards), rice, then soya and corn on the black soils. A paradoxical land, due to past and remaining poverty, even after the abolition of slavery that was accompanied by a collapse in the former economy and densification of the African-American population that was at last free and willing to cultivate their own land plots, but irremediably poor; *tenancy* (a form of metayage on lands belonging to large landowners) was a new form of slavery. The deltaic landscape has shaped a society that is socially significantly contrasted: large Euro-American owners and large numbers of African-American slaves contrasting a society that had an original culture and which was relatively homogeneous in terms of diversity across the hundreds of kilometers of the lower valley. It had become used to the recurring devastation of high water levels[9] and floods which put all their efforts into question, to malaria, to hookworm and to various fevers in an exhausting climatic context. A country of masculine violence in an environment with a tendency for isolation, as seen in the literature ranging from Mark Twain to Ernest J. Gaines (*A Gathering of Old Men*), with more recent Tim Gautreaux (*The Missing* and *The Clearing*) or even Tom Cooper (*The Marauders*). In addition, it is a land of militant evangelism, motivated by women who were invested in the cause of maintaining a certain level of morality in a land riddled with all kinds of excesses, violence and a paternalistic society originating from slavery.

> "The same watercourses that enriched the country and which served as arteries for trade and travel before the arrival of the railways and motorized transport have also inhibited the settlement of mankind, caused isolation and death, and regularly ravaged the landscape. Waterfront cities have experienced routine collapses which have swallowed up roads and their buildings; in certain cases, entire towns have disappeared into the water" [BAR 97].

Having become, between the Civil War and World War II, an area of cotton monoculture, the Arkansas Delta is today an open country, a stronghold of agricultural business, from widespread mechanization to GMOs, from irrigation to pesticides and water pollution. In its recent history, it has experienced great changes; strong demographic growth at the end of the 19th Century, a reduction in population from the 1920s onwards, with the rise in mechanization then the cotton crisis (weevils and falling prices); the population of African-Americans in the region then became dramatically lower, a persistent "sub-culture in mass society", and lastly the paradox "of a people who are amongst the poorest in the world, living on one of the richest lands" [COB 94].

With the loss of 17% of its inhabitants between 2000–2010, Greenville (MS), "the heart and soul of the delta", is the epitome of a small town that is losing momentum at the heart of the rural region in crisis. The 12 counties in the Delta have lost between 50–75% of their population since 1940. The closure of factories and fish farms has followed the loss of rural jobs.

In the continuity of the slave economy, low salaries, a low level of education and obsolete technology do nothing to prepare this region for international competition; only agricultural productivity and profound ecological impoverishment. This reduction in dynamism and in the population of aged counties will allow the delta's crisis to be put into perspective [FAS 13a, FAS 13b].

9 The 1927 high water levels and flood led to hundreds of thousands of African-American inhabitants leaving for the north of the United States [BAR 97].

The "deltas" in Arkansas and the Mississippi were named during a time period in which the similarities that they had with the Mississippi Delta had become more important than the differences, in particular the presence of salt and progradation into the sea; for the rest, the landscape in which the inhabitants lived was without doubt very close in the two types of zones. The situation today is no longer the same, because the beautiful alluvial forest has disappeared, the bayous are no longer connected to the river and their landscapes have been lost.

4.4. The Mississippi Delta *stricto sensu*: a natural zone in crisis

4.4.1. *Flow and landscape dynamics*

The "Mississippi River Delta" is the official denomination of a vast zone in a classic triangular form, encompassing 25,000 km^2. The addition of "River" allows us to specify that we are entering a zone constructed by the river at its mouths, independent of the local names upstream. The small town of Angola pinpoints the entrance to the delta itself, even though this distinction at the heart of the immense alluvial zone has quite an arbitrary aspect to it. It is at this point that the river loses part of its water, on the right bank, as it joins the Red River and its extension, the Atchafalaya.

During the Pleistocene, the Mississippi gouged out a valley 115–135 m deep and it exits into a submarine canyon that cuts into the continental shelf. The valley began to be infilled around 12,000 years BP and the delta has been formed since stabilization of the sea level around 7,000 years BP; the delta has risen on average by less than 1 mm/year since the end of eustatic rise. Between 2,780–3,450 billion tons of sediments have been deposited between Memphis and the Gulf of Mexico in the last 12,000 years, giving a cumulation rate of 230–290 Mt/year (corresponding to the transported tonnage of 400–500 Mt/year). The sediments have prograded downstream from Baton Rouge for the last 4,000 years, benefitting deltaic branches. With this purpose in mind, it is legitimate to consider that the geological delta begins at Memphis and that the deltaic form *stricto sensu* ends only in Louisiana. The change in deltaic landscape, which is largely explained in terms of the relative rise in the sea level, fundamentally depends on the sedimentary input of the Mississippi [BLU 09]. During its history in the Holocene period and its diffluences, the Mississippi has constructed a multi-lobed delta.

Branches and lobes of the delta complex	Dates before the present day (various authors)	Dates [DAY 07]
Maringouin-Salé-Cypremort	7,500–5,000	4,600
Teche	5,500–3,500	3,500–2,800
St. Bernard	4,000–2,000	2,800–1,000
Lafourche	2,500–500	1,000–300
Plaquemines-Balize	1,300–0	750–500
Atchafalaya-Wax Lake	500–0	

Table 4.1. *The successive deltaic complexes of the Mississippi Delta (using traditional concepts and according to [DAY 07])*

A new deltaic complex was created approximately every 1,000–1,500 years, with the most recent created by avulsion of the Atchafalaya, which is controlled by the US Corps of Engineers. The Plaquemines-Balize lobe controls recent wet environments which are adjacent to it. On the left bank of the Mississippi, to the northwest of New Orleans, there is the vast Maurepas and Pontchartrain Lakes (Table 4.1). The oldest lobes are subject to actions from the sea, which shape their coastline into islands and sandy beaches after the construction of cheniers that are now surrounded by land; barriers islands and cheniers protected the delta from marine intrusions. When the first Europeans arrived, several distributaries were active.

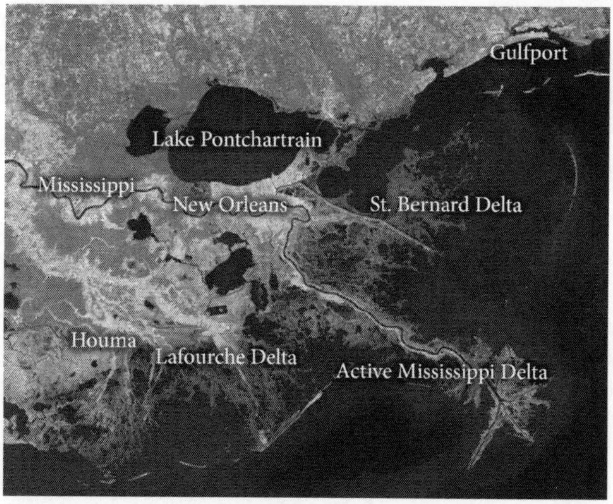

Figure 4.1. *The Mississippi channel at New Orleans, Lake Pontchartrain to the north of the city, the current Plaquemines-Balize lobe; further south, the light band of the old Lafourche lobe. This view is an overlay of USGS Landsat 5 images from October 3 and November 11, 2011. For a color version of this figure, see www.iste.co.uk/bravard/sedimentary2.zip*

The Mississippi Delta is not of a unique type. As part of its operation, as much fluvial as coastal, the delta undergoes forcing and pulsing events. These are changes in the course of the river, floods, storms and tides. Contrary to a preconceived idea, sediments of marine origin that contribute to current compensation for subsidence come more from flooding by high tides than from inputs during frontal depressions, tropical storms and hurricanes [MOR 02]. On a small scale, the delta is river-dominated in its active mouths; however, the action of tides and waves prevails over the sections of coast that are no longer supplied with sand by the active mouths. At what rhythm? The current processes are of a very modest size – the Gulf imposes a microtidal regime and very low wave energy – but the frontal systems and the hurricanes can have significant effects due to the energy that they give off [GEO 05]. In addition, the reduction of sedimentary inputs of fluvial origin unbalances the system and is instead beneficial to the coastal redistribution of sediments.

Dyking works on the river are primarily responsible for the current problems, insofar as they prevent the fluvial floods from distributing their fine sediments at the surface of the area that had previously been flooded. The construction works carried out since the middle of the 19th Century to protect communities, economic infrastructure and harvests have made normal construction of the deltaic plain in the form of marshes impossible, through inputs from fluvial floods that spread out, leading to breaches in distributaries that supply the former lobes. In its natural state, each crevasse splay* covered on average a surface area of 1,500 km^2 and the breach that was responsible had a limited life expectancy of less than a century.

Owing to the construction of levees, the floodplain area that was likely to withhold sediments between Cairo and the downstream part of the delta was reduced from 89,600 km^2 in its natural state to only 7,000 km^2 today. In the period between 1850–1927, ruptures of deltaic levees occurred in one-third of years during flooding. Civil engineering works have reduced the number of natural access routes to the marshes (the crevasses have been brought under control) and the flood flows have been concentrated with respect to the original state. As a result, there has been in particular a complete halt to inputs to the Lafourche bayou, which concentrated 15% of the sedimentary circulation around 1850; it is true that the Atchafalaya experienced a rise in its proportion of the total load from 12–30% (the Mississippi in New Orleans still transports 70% of the load, compared to 73% in 1850) [KES 03].

As a result of these modifications, and with the exception of the two recent lobes under construction, the delta is considered to be in a transgressive phase, meaning a phase of rising sea levels and of land loss to the ocean. Since the 1930s, the delta has lost a land area of more than 4,900 km^2, in other words the size of a football pitch for each hour that goes by. The delta is sinking, posing a threat to the population, the uses and wildlife.

4.4.2. *The Atchafalaya and its deltaic lobes*

Near Fort Adams, but on the right bank of the river, Atchafalaya is the name given to the lower course of the Red River. The Atchafalaya flow rate has benefitted from partial defluviation of the Mississippi, which took place in the 16th Century. This branch now transports approximately 25% of the flow of the Mississippi during the spring maximum, in addition to the total flow rate of the Red River. The flow rate coming from the Mississippi is controlled by the Old River Control Project (1963), constructed to eliminate the risk of total capture of the Mississippi by the Atchafalaya and to conserve the trading economy of New Orleans and the lower Mississippi corridor.

The main criticism of the US Corps of Engineers, already criticized for certain aspects of its river management, involves the distribution of the flow rate between the Atchafalaya and the main channel that passes in front of New Orleans, within the delta. The Old River Control Structure (1963), extended by the Morganza Floodway, means that the Atchafalaya can evacuate more than 25,000 m^3/s during a flood (without considering the flow rate of the Red River and that from the two spillways placed downstream from the Old River), in total more than the Mississippi river in front of New Orleans. This inversion of the importance of the two branches is probably the reason for sand deposition in the river downstream from New Orleans and poses the threat that the Atchafalaya (which is much shorter, flows much faster, is straighter, and the incision of whose bed is increasing) should become the main channel, in such a way that the primary problem is now to conserve the Mississippi in its bed and to avoid a sea arm ascending to Baton Rouge along the Atchafalaya following a disastrous overflow, making use of the over-incised and easily exploited channel [REU 04].

Between 1980–2009, the average annual flow rate of the Atchafalaya remained reasonably constant at 190 km^3, despite interannual fluctuations. On the contrary, the sediment load moving past the Atchafalaya was reduced from 66 Mt/year to 41 Mt/year during the same period, owing to the reduction in the Mississippi's sediment load and that of the Red River, which is itself equipped with dams with locks to allow navigation.

The Atchafalaya, a river with a sandy bottom, underwent large amounts of work to remove logjams in 1833 and in the 1870s, as well as dredging in the 20th Century. It has its own floodways which supply marshes, vast expansion basins of floods over an area of more than 5,500 km^2; in addition, the small town of Morgan City is protected by a circular dyke and the Wax Lake artificial channel was dug towards the sea between 1938–1941 to act in parallel to the Atchafalaya. The maze of old channels of the Red River/Atchafalaya has filled in the lakes and marshes of the Cajun Country alluvial plain and constructed a double lobe on the coast; this lobe emerged after the 1973 flood and extends along two subaqueous deltas in the Atchafalaya Bay (Figure 4.2). The delta has prograded by more than 200 km^2, but this growth is reduced due to the reduction in fluvial inputs and the variability in the intensity of tropical cyclones [MOS 16].

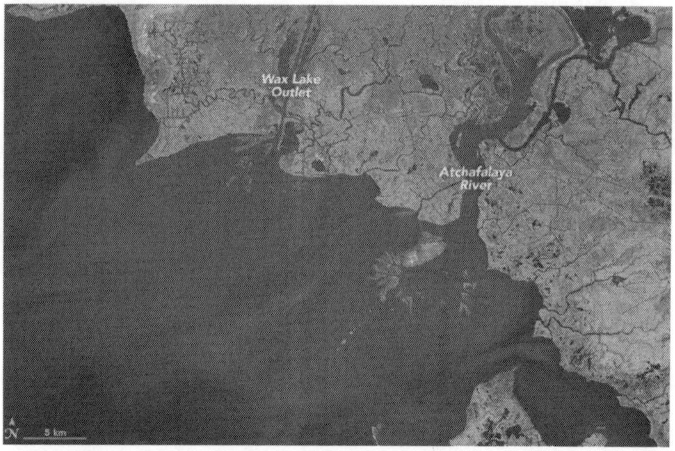

a)

b)

Figure 4.2. *Despite the reduction in its load between 1980–2009, the Atchafalaya has conserved its capacity to construct two small lobes at its mouths. a) Situation in 1984 and b) situation in 2016 (images taken during the autumn low-water season). Combination of Thematic Mapper on Landsat 5, Enhanced Thematic Mapper Plus on Landsat 7 and Operational Land Imager on Landsat 8. (Source: NASA Earth Observatory – Jesse Allen). For a color version of this figure, see www.iste.co.uk/bravard/sedimentary2.zip*

4.4.3. *The conversion of delta marshes into free water and coastal regression*

Today, exactly half of the surface of the delta is above the sea level, with the other half below it. Delta sinking and coastal regression processes are related to subsidence. This is accentuated by the eustatic rise in the sea level, a gradual process, and all this at a pace approaching 1 cm/year. The regression of the marshes and correlative extension of areas of open water in the deltaic plain have reduced the roughness* of the delta and accentuated the fragility of this area to destructive hurricanes that sweep unrestrictedly across it (Figure 4.3).

Figure 4.3. *Erosion of the deltaic coast threatens housing (Grand Isle barrier beach) in 1992; rip-rap has been put in place. (Source: J.-P. Bravard). For a color version of this figure, see www.iste.co.uk/bravard/sedimentary2.zip*

First, let us look at the facts. The plain and the deltaic coast are experiencing an accelerated evolution of their wet areas. These were constructed over the course of the centuries by the formation of peat on fine mineral sediments deposited between the lobes under construction and on the vast muddy stretches of the chenier plain located at the far end of the coastline, to the west of the historical lobes. Conversion of vast areas of open water marshes was rapid, with a loss of nearly 3,000 km^2 over the period 1956–2004 and 4,820 km^2 over the period 1930–2015; this loss makes up 80% of the reduction in wet zones recorded everywhere in the United States. This dynamic had slowed down since 1978 and even more since 1990, even though the erosion affects riparian coasts. However, the creation of new, small, marshes is still active and has become the most significant form of evolution within the delta, in particular for the Pontchartrain, Breton and Plaquemines parishes. Changes have been observed in the areas of freshwater "floating marshes", which were ripped apart by winds during Hurricane Andrew in 1992 and Hurricane Katrina in 2005. In another change, the coast of the Gulf of Mexico and the sandy cheniers are receding significantly, in places where lakes are highly developed and the wind action is strong, which is compatible with local advances, such as in the mudflats of the Atchafalaya and the Wax Lake Outlet.

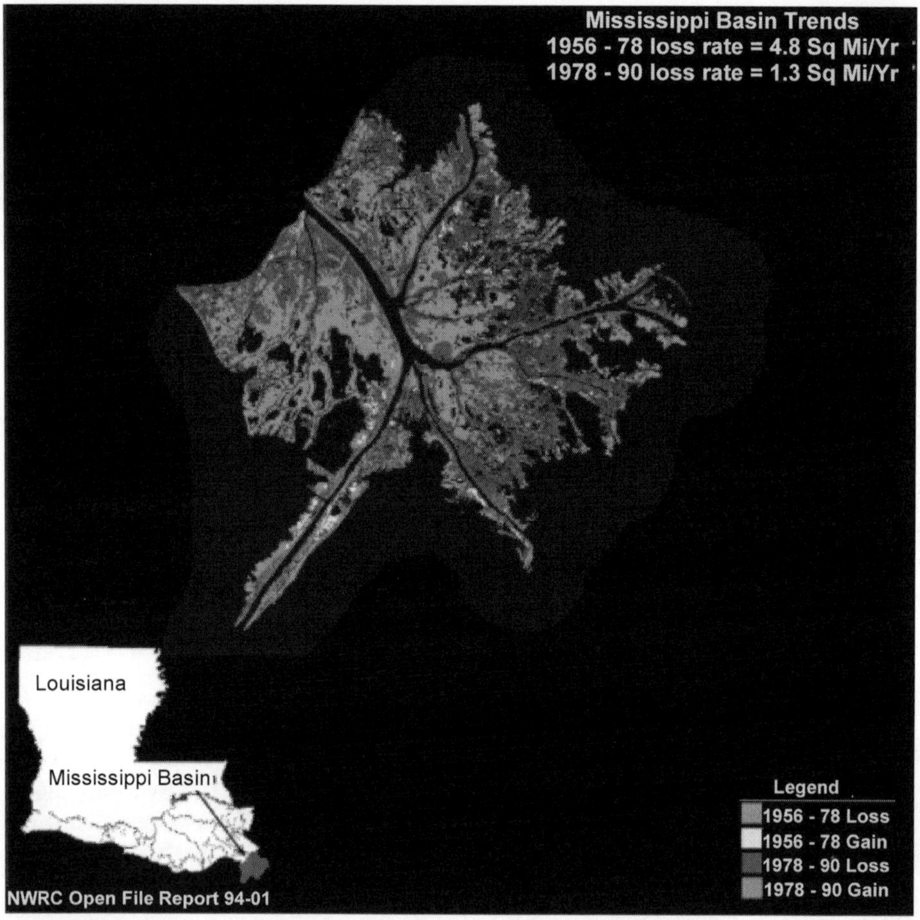

Figure 4.4. *The area of the lobe under construction downstream from the Mississippi was reduced by 12.4 km²/year over the period 1956–1978, at a rate of 3.4 km²/year for the period 1978–1990. The distributaries increase their area, but the older terrains are getting smaller due to subsidence and the rise in sea levels [BAR 94]. For a color version of this figure, see www.iste.co.uk/bravard/sedimentary2.zip*

Subsidence of the delta is accentuated in the areas located between the distributaries of the deltaic plain. Owing to compaction and loss of water in the sediments, subsidence was slow from 1947 to the middle of the 1960s (less than 5 mm/year), then it doubled in speed from the 1960s to the beginning of the 1990s (approximately 10 mm/year), before renewing, it would appear, a natural pace from then. It is admissible, simply by chronology of the facts, that the sectors where the loss of marshes has been the greatest coincide with those where the continental extraction of hydrocarbons (oil and gas) has been the most intense, where there is sufficient reduction in volumes of pores to explain the phenomenon. Subsidence is not only due to compaction. Without a doubt, extraction activated faults (Figure 4.4).

It is also important to not neglect the effect of the intensive drainage of Greater New Orleans by the US Corps of Engineers in order to convert the marshes into areas “to be developed”. The waters were released into Lake Pontchartrain. Intended to reduce the level of the water table, these large-scale pumping and drainage works have accelerated compaction of the adjacent organic soils over a height that locally reaches 2–2.5 m. In the 20th Century, more than 15,000 km of artificial channels have been built across the delta to accompany the extraction and exportation of hydrocarbons. The effect has been an alteration of the hydrology due to concentration of the flow, reduction in run-off at the surface of the marshes and a circulation of nutrients at the surface of the deltaic plain, as well as facilitation of salt penetration far into the marshes, in such a way that the freshwater vegetation dies off and the peatlands fall into even worse condition. The Mississippi River–Gulf Outlet, dug in 1963 to the north of the Mississippi channel between New Orleans and the Gulf to avoid the main mouth, has been made responsible for the death of thousands of hectares of forest in the freshwater marshes. The “accelerated historical subsidence” has especially affected the southwest of the deltaic plain, by creating considerable “accommodation volumes”, which are taken into account in recent restoration projects [BER 13].

These results, albeit recent, are not in phase with those provided by another study which used nearly 1,200 topographical markers from 1922–1995 and applied an interpolation method. The average subsidence, defined as the compaction of the (natural or accelerated) soil was 9.4 mm/year over the period in question; subsidence is greater in the districts of Orleans, Jefferson, Terrebonne and Plaquemines. Consequently, annual growth of the areas threatened by flooding (since they are located below sea level) increased from 44.3 km^2/year around 1920 to 241 km^2/year around 1950, as much for marshes as for land that is currently agricultural, before stabilizing around this figure [ZOU 15].

The loss of surface area of marshes is not only due to accelerated natural subsidence through the extraction of hydrocarbons in certain places. Multiplication of introduced wild fauna, and some large construction projects and hurricanes, have seriously affected the marsh ecosystems. Invasive vegetative species have a disastrous impact; for example, water hyacinths, an ornamental species introduced in 1884 to New Orleans during an international exposition, within a few years invaded and obstructed the navigation system of the lower and upper Mississippi. Wild boars originating from Europe (*Sus scrofa*) were introduced in the 16th Century; these graze aggressively and, other than the damage they cause, carry diseases that can affect humans (brucellosis, herpes, classical swine fever) [HOL 11][10]. Rodents with rapid reproduction were introduced to help develop the fur trade. They dig tunnels in the banks, feed on the roots of marsh plants and devour young cypress specimens. The coypu (*Myocastor coypus*), for example, imported from Argentina into California, was introduced to Louisiana in the 1940s. It was a lucrative hunting target from 1960–1990, but the fur market has collapsed since then, in such a way that the species has proliferated and it was considered necessary to control it with a Coastwide Nutria Control Program set up in 2002 to protect the marshes (more than 2.5 million coypus have been eliminated since 2002). Simply due to their consumption of vegetation, the 2 million coypus in the delta (the current population) subtract the equivalent of 45,000 tons of carbon each year), preventing the regeneration of the forest of bald cypress trees and reducing the marsh areas. In addition, negative effects of storms and hurricanes on the population of coypus, and the growth of alligator populations, contribute to regulation by predators, which do, however, remain to be proved. Lastly, in 2010, the disaster of the Macondo offshore oil platform, exploited by the company BP, led to 800,000 m^3 of oil being spilled from the Deepwater Horizon underground reservoir, which then polluted the coasts of Louisiana and destroyed part of their ecosystems.

4.5. Hurricanes and their effects on the Mississippi Delta

4.5.1. *Hurricane Katrina*

Up until now, we have looked at the fluvial component of risk, considering that the floods formed in the Mississippi were the threats posed to the valley, the delta and New Orleans. Yet the delta is subject to other constraints created by meteorological situations that are particular to the Gulf of Mexico. The north coast of the Gulf is crossed by "tropical depressions" (winds up to 60 km/h), "tropical storms" (winds from 60–120 km/h) and lastly "cyclones", here named hurricanes[11] (winds at speeds greater than 120 km/h). The first well-documented hurricanes appear to be the 1901 storm, then the 1955 (Janet), 1957 (Audrey) and

10 Website of the Louisiana Department of Wildlife and Fisheries: http//:www.wlf.louisiana.gov.

11 The term "hurricane" comes from the Spanish word, "huracan", recorded since the beginning of the 16th Century (CNRTL – Centre national de ressources textuelles et lexicales).

1961 (Carla) storms. Hurricane Betsy alerted the authorities to this risk in 1965 and meant that the new Naval Air Station Meridian (NAS Meridian) became the fallback base for airplanes from the aeronautical forces in the United States, too threatened by hurricanes on the coast, especially by the 1968 (Gladys) and 1969 (Camille) hurricanes. Recent climatological research tends to consider that the intensity of Atlantic cyclones and hurricanes has increased since the 1980s and that the area they cover is extending further north; storm surges in the sea level are associated with this and present a danger at high tide, since they are generally about 1 meter in height. This evolution is generally attributed to global warming, in particular to the warming of oceans and of the Gulf.

Cyclone Katrina (August 23–31, 2005), with a peak wind speed of 280 km/h, is considered to be the third largest in the region for the intensity factor; it was classed as a category 5 hurricane once it passed over the warm waters of the Gulf of Mexico[12]. Despite the evacuation order given on August 27, there were over 1,460 human deaths and the induced effects of the disaster were considerable, with losses evaluated at 108 billion dollars overall for all areas from Florida to Texas[13]. The hurricane opened more than 50 breaches in the defenses of the Mississippi Delta and flooded 80% of coastal cities. The rise in level of Lake Pontchartrain under the effect of the north wind (a storm surge), to which 2 m-high waves were added, was the primary cause of the disaster. 50% of the breaches and human losses due to Hurricane Katrina took place in the Greater New Orleans area, 80% of which was flooded due to the rupture of the dykes placed alongside the navigation channels. In addition, the Mississippi River–Gulf Outlet, wide open to the Gulf, was in a position to channel the winds and aggravate the rise in water level of Lake Pontchartrain, as was also the case for other canals.

Despite the warnings given to the urban authorities and to the US Corps, a plan made in 1965 (the Hurricane Protection Project), itself judged to be deficient, was far from being completed or was carried out to an insufficient standard. Hurricane Katrina is considered by some to be the worst disaster caused by engineering in American history; the responsibility of the US Corps of Engineers was directly called into question. It had created a badly coordinated system which did not account for the reality of subsidence [MAR 08]. In addition, approval of the National Flood Insurance Program (1968), which provided protection for the new, exposed, constructions in the floodplain, and the confidence placed until 2005 in the quality of structural protections, had dissuaded the authorities from implementing a policy to reduce

12 The section about Katrina and New Orleans is largely inspired by the following Wikipedia pages: "Hurricane Katrina", "Levee failures in Greater New Orleans" and "Effects of Hurricane Katrina in New Orleans".

13 In September 2005, Hurricane Rita hit Louisiana and forced another evacuation of New Orleans. It is classed fifth in the hierarchy of hurricanes occurring in the Atlantic area since 1893.

vulnerability – an additional factor to take into account is that management of the risk was unsound at various levels. The drainage works (see previously) meant that approximately 50% of the urban area of the agglomeration found itself constructed below sea level when the hurricane arrived, whereas vast areas located above sea level had remained free of any residential construction. Subsidence partly explains the losses. It is largely responsible for the increase in population that was living below sea level at the end of the 20th Century (poor and for the most part African-American), with an insidious and severe degradation of the situation [CAM 07]. An important piece of information is that Hurricane Katrina hit a city in crisis, a city in economic and demographic decline since the 1960s; a city that was partly undeveloped and which implemented bad management of its property, and in which the urban built environment was too weak. The city "was emptying but was not getting smaller" in size, as good protections would require; the population thus reduced by 10% in 10 years within the perimeter that was supposed to be protected (a little less than 820,000 inhabitants in 2010) [ZAN 12, ZAN 13].

4.5.2. *What does the future hold for New Orleans?*

4.5.2.1. *What should the response be to this crisis?*

Opinions of the inhabitants and experts are divided between "abandonists", "maintainers" and, on central ground, "concessionists"; the latter are supporters of shrinking the area of the city to reduce its exposure to risk, while conserving its heritage and its functions. The maintainers' reasoning, largely meaning allowing nature to take its course, has prevailed in the end, because it was the least conflictual, if not reasonable; this laissez-faire approach has been chosen by urban authorities and the federal government. According to G.E. Galloway, "the concept of management of flood risk is still in its infancy in the United States and the federal policy has created a bias in the way the environmental component is taken into account" [GAL 08]. The New Orleans Urban Planning Committee did in fact admit in 2006 to the logic behind putting more modern methods into practice. The Unified New Orleans Plan (2007) should in principle and in the long-term remedy these errors and these deficiencies [MAR 08]. Also, part of the population has relocated to areas above sea level. In other words, exceptionalism, arising from colonialist roots and resting on the delta's cultural identity, has prevailed over rational Americanism which advocated abandonment. Why has it prevailed and has it obtained the subsidies required for reconstruction? This is because it is the basis of community pride in New Orleans, a mixed city haunted by its own history[14]. A choice that

14 On this subject, a suggestion would be to watch the dramatized series *Treme* by D. Simon and E. Overmyer, broadcast in the United States on HBO between 2010–2013. Faubourg Treme is formerly a neighburhood of non-slave African-Americans, near to the French Quarter.

sometimes has a very high human and financial cost. This is the dilemma that this unique city is facing [CAM 08].

Figure 4.5. *Aerial view of New Orleans in 1992. The French Quarter with low houses constructed in an organized layout is on the right of the photo (to the north). The river is flowing towards the camera. (Source: J.-P. Bravard). For a color version of this figure, see www.iste.co.uk/bravard/sedimentary2.zip*

4.5.2.2. *Should we aggrade the city?*

Certain researchers tackle the subject of the future of the city by looking for analogies in the past that would be, according to them, more reliable than models predicting the behavior of individuals. Historians consider that risk-taking inherent in individuals is highly diverse and defies regulatory ideas based on scientific concepts of resilience and vulnerability. The singularity of human adjustment relies on the better incorporation of resources and adaptative capacities. Analogies from the history of the United States have been found in Chicago, when a cholera and typhoid epidemic hit the city while it was flooded when sewer effluents were released in 1854. A new network was positioned and buried in systematic infill of a thickness of 2.40 m. The city of Galveston was protected by a wall over 5 m high and then infilled with sand up to the top of the wall; the buildings were aggraded to the same height.

The idea of raising the elevation of New Orleans by sealing off dyked marshy compartments dates back to the middle of the 19th Century, but nothing was carried out until the 1930s, behind a defense barrier set up facing Lake Pontchartrain.

Today, the principle of increasing the relative elevation of New Orleans focuses on dredging sediments from Lake Pontchartrain and the fluvial channel to widen the levee and block off the low-lying land before constructing on them. This method would create a patchwork of flood basins and space for houses in an elevated position. Hurricane Katrina was certainly an event that was likely to begin a movement in society, but the city and Louisiana have been through an economic crisis, even though subsidence risks to discourage initiatives of this kind in favor of renaturation projects. However, it seems that a project of this kind will above all find itself up against a tired and skeptical population and inhabitants of the non-urban margins who are hostile to any project that is likely to push water back onto their territory, as shown by history in the 1960s. While it is not possible to replicate these analogies, they evoke the wisdom of making prudent decisions [COL 18].

4.5.2.3. *The delta's inhabitants*

The crisis caused by Hurricane Katrina came in the context of economic and demographic depression that the disasters made significantly more fragile, and it became a physical framework for high levels of spatial instability. One of the features characterizing the delta is a fundamental opposition between the urban communities in the New Orleans Metropolitan area, economically developed and protected by levees (but which remain vulnerable to storm surges created by hurricanes), and all the floodable and highly vulnerable neighborhoods. The evolution in progress has had harmful effects on the human and cultural environment [GRA 05]. The marsh was both an ecosystem and a way of life for deeply rooted local inhabitants who had built up a "sense of place" with the area in which they lived. This culture was formed from the combination of multiple crops, added to the old Native American foundations (oyster harvesting brought by the Croats, shrimp drying from the Chinese, sugar production from the West Indies); a great number of cultures with knowledge of living around water made their contributions. More so than the others, the Cajuns, who arrived in 1755, defined this lifestyle by combining fishing, hunting, furs and agriculture, which they had been familiar with in Acadia. An entire area encompassing natural aspects (diversified resources), economic aspects and social aspects (Cajun culture) is threatened by water. Resilient to hurricanes, this population is threatened by a slow environmental disaster, with a reduction in marshy surfaces and a rise in sea levels. Spatial instability is a hazard that indiscriminately affects groups that are sometimes very diverse. For instance, the Filipino fishing community in Saint-Malo (Biloxi Marsh, to the east of the St. Bernard neighborhood), established at the end of the 18th Century, was wiped out by the 1915 hurricane. The vulnerable communities are largely formed by African Americans. Let us consider, for example, the community of Fazendeville (St. Bernard neighborhood) which was dismantled abruptly in 1964 to liberate and protect the site of a Civil War battleground; the community, relocated nearby, was again affected in 2005. One of the tragedies that these delta populations experience, in particular small scattered minorities, is the inadequacy politicians in the federal government, who constantly choose to relocate and re-establish them. These communities, generally rural and poor, have in fact been "dislocated" [DAL 14]. Hence, considerations of post-Katrina territorial reconstitution are made in terms of urban adaptation to risk. Since the response from federal authorities has been insufficient, and with responsibility removed from the communities themselves, the situation will be played out among second-order actors, namely the residents' households themselves [ZAN 13].

4.6. Sediments in the Mississippi and equilibrium of the delta

4.6.1. *Simply a reduction in inputs or a sediment deficit?*

Over the period 1976–2006, the tonnage carried in suspension by the river in the delta is estimated to have been 205 Mt per year, including 136 Mt for the Mississippi passing in front of New Orleans and 69 Mt for the Atchafalaya. Input to the delta in terms of sediments is today therefore limited with respect to what it was in the 19th Century. An additional fact appeared, following the Great Flood of 1993 and caused by very high rainfall on the upper Mississippi basin and on the lower Missouri[15]. The 1993 flood has had the long-term effect of abruptly reducing sedimentary transfers, which was felt at least up to 2010. The hypothesis is that the flood has caused flushing on a massive scale of sediments stored in the fluvial channel and which could easily be mobilized, whereas the protection policy of land in the basin and the construction of structures in the river limit the upstream and lateral inputs; sediments that are currently in transit are probably trapped in the structures that were purged in 1993. The question is whether or not this trend is going to be long-lasting [HOR 10].

The cause is frequently attributed to the reduction in inputs coming from the upper basin, in accordance with traditional approaches to the issue. But is this really the cause of the delta's current problems? For certain researchers, the origin of the problems is more to do with disconnection of the deltaic plain and of the main channel of the Mississippi. We note that the lower Mississippi is surrounded almost continuously by high levees spaced 1 km apart on average, whereas the flows of the Atchafalaya are freer to disperse themselves over an area of 4,000 km^2. Resulting from this, the deltaic plain of the Atchafalaya is aggrading at a rate of 44 mm/year and its mouths are prograding significantly. The solution lies in the organization of transfers from the main fluvial channel towards its edges by the controlled overflows and by diversions [XU 14].

15 The 1993 flood is the highest recorded since 1931 (20,600 m^3/s at Tarbert Landing), which attributes it a return period of 50–100 years.

4.6.2. *The rise in sea levels and climate change*

Predictions made by the Intergovernmental Panel for Climate Change (IPCC) agree on an acceleration of the rise in sea levels. Following an average of 1.7 mm/year in the 20th Century, valid for the stable regions of the world, but with 3 mm/year since 1993, the rate of rise could increase to 4 mm/year in 2100. Maintaining a sediment deficit which would be, at the scale of the delta, in the range 3–9 Gt/year, would unavoidably cause it to be flooded. The relative rise in sea levels would cause additional losses of between 10,000 and 13,500 km^2 in the delta before 2100, at a rhythm which is three time greater than that of construction of the deltaic plain; compensation for these losses would require a total estimated volume of between 18–24 Gt of sediments of continental origin, within the same period, which cannot be envisaged if the current inputs are taken into account [BLU 09]. The seriousness of the current situation lies in the fact that the delta and the coast could in the future be subject to more violent storms and to an accelerated rise in sea levels due to a combination of the greenhouse effect (increased due to the consumption of fossil fuels) and to the local extraction of hydrocarbons which lie under it; some see the irony in the fact that the delta's oil wealth is in itself leading to its loss [COS 06].

Many researchers are asking questions about the future behavior of the marshes now that they are facing a relative rise in the sea level. Coastal areas that are expected to recede the most are located in the active lobe and in the Lafourche lobe [BAR 03]. The quantities of sediments that would be required to restore the marshy landscapes to their 1956 state, a reference date, as much on the surface as in elevated areas, have led to a calculation of the "accommodated volumes" following extraction of hydrocarbons. Diachronic research has laid down the basis for an understanding of the resilience of these environments and the foundations for their restoration based on predictive models [BER 13].

4.6.3. *Reconstruction of the marshes*

The United States government has spent 85 million euros repairing New Orleans and the Gulf coast, and this figure is going to increase, because the federal state intends to protect a delta that contains the country's primary port which provides a third of its oil and gas. The option that was very quickly selected by the US Corps was to reconstruct levees in exactly the same way, or nearly so, in order to reconstruct an economic tool as a priority. Yet many experts suggested restoring the marshes instead, using the forces of nature rather than traditional engineering techniques. This would be the common-sense solution, insofar as the delta is one of the largest fishing regions in the United States and a major bird migration site. In addition, because the open and wooded marshes form a natural barrier against storms, whereas the stretches of open water accentuate wave action (by increasing the fetch*) and storm surges*. Inversely, the presence of an 80 km band of marsh or shallow water in front of New Orleans would reduce the storm surge facing levees by more than 3.5 m in height. Specific studies have demonstrated that the propagation of surges (these violent intrusions of water from the sea that occur during a storm) across marshes reduced the relative height of water by 5–10 cm/km of intact marsh [DAY 07]. With this in mind and with regard to many other aspects, Louisiana's wet areas produce service of a value estimated to be 800 euros/ha/year just for natural defense of the coastlines (the cost of destruction of 4,800 km^2 of marshes is 5.1 billion euros/year). We understand the interest of a prevention policy if we consider that restoration of the marshes would cost a mere 21.3 billion euros [COS 06].

4.6.4. *Sedimentary management of deltaic branches and the future of the marshes*

A rapid and effective method would be to dredge the channels in the marshes with the aim of transporting mud and sand towards areas that are to be reconstructed as a priority. This method of engineering is backed even more due to the fact that dredging has for a long time been an intensive practice carried out in the name of navigation, because researchers have been able to demonstrate its effectiveness in restoring stretches of salt marshes of *Spartina alterniflora* affected by the drought across hundreds of km^2. Aggradation of the surface of a marsh is beneficial for vegetation and for the construction of organic soil; a conclusion that has led to prohibitions being lifted with regard to this costly and heavy method. However, if USACE dredges on average 60 Mt/year, of which 20 Mt would truly be useable, restoration carried out in this way would have a very high cost. The method is selected above all because its localized results are seen very rapidly [KHA 15].

This is the reason why the best choice seems to be to open diversions that transfer loaded floodwaters towards low-lying zones. This design advocates to deviate, with diversions located further upstream of the loaded waters, in the same way as the processes at work in living deltas, towards partly submerged or emerging zones. The idea is to imitate natural operation (prior to the colonial period, at least), which involved forcing under the influence of high fluvial or marine energy levels; this was capable of creating new lobes and opening breaches in a network of fluvial levees and barrier islands surrounding sedimentation basins that are protected from marine intrusions. This method also leads to an increase in the organic contribution to the aggradation of marshes, at least to an increased efficiency of trapping the sediments that are transiting through the marshes, ponds, tidal channels and shallow bays, at least where storms are likely to remobilize sediments and to transport them towards the interior of the delta [BLU 09]. By reintroducing loaded fluvial water, deposits aggrade the ground, freshwater reduces salinity, iron bonds with sulfites lose their toxicity, and organic soils become able to re-establish themselves. The Mississippi has opened up a thousand natural breaches between 1850–1927, in such a way that behind the levees, splays occurred successively in a wide band whose surface dipped slightly towards the marshes. The only breach in Caernarvon, voluntarily opened in 1927 downstream from New Orleans, formed a spread covering 130 km^2 within one month. A breach that was voluntarily opened in 2002 opposite New Orleans (Davis Pond River) produced effects comparable to those of natural breaches [DAY 16a].

Thus, things could work out in terms of the re-equilibrium of masses at the scale of the active delta, on the condition, however, that sediment diversion sites are carefully and successfully chosen. The Mississippi has high flow rates in spring (which exceed 30,000 m^3/s); the tides, which are felt up to 50 km in the in-flood river, cannot inverse the current. The diversions would therefore operate above all in spring. Models have led to the study of projects to deviate floodwaters loaded with sediments to the south of New Orleans; the priority sites would benefit from 45% of the river's load downstream from the city (or from 25% of the total load as it is measured upstream of the Atchafalaya diffluence). Enough to reconstruct between 700–900 km^2 of land before 2110, certainly, but this increase in area (rather hypothetical it must be said) corresponds to less than 10% of land which, according to other models, will be submerged in 2100.

Generally, diversions would also have the positive effect of retaining, within the deltaic plain, a significant fraction of the nitrates provided in abundance by the Mississippi basin due to overfertilization of crops, agricultural drainage and inputs from urban sources. In the absence of a reduction at source of the flows of nitrates and compensation for the loss of 14 Mha in wet areas in the basin since the 1780s, this method would have the positive effect (at least theoretically) of reducing the hypoxic* "dead zone" which extends into the Gulf of Mexico in summer [MIT 01]. Restoration of marshes would also reduce the drastic carbon losses caused by hurricane damage, as was the case for Katrina and Rita. The loss due to these two hurricanes was 15 Mt of carbon compared to a net storage gain of 1 Mt/year on average in the marshes (over a total of losses due to this factor of nearly 150 Mt since the 1930s); at the expected pace of reconstruction, around 50 years would be necessary to more or less recover the amount of carbon lost by the delta following the two hurricanes [DEL 11].

4.6.5. *Coastal protection plan*

The scientific results of the project Coast 2050, based on a retroactive analysis, and the innovative measures have been incorporated in the 2017 Coastal Master

Plan, which envisages expenditure of more than 50 billion dollars over 50 years (until 2067). The plan relates to the following objectives [CPR 17] (Table 4.2).

Nature of the project	Cost (billions of $)
Construction and maintenance of the ground for an area of 800 km^2	
Creation of marshes by dredging	18
Diversion of laden waters	7
Other actions	2
Reduction of the flood risk with reduction of losses by 8.3 billion/year	
Structural measures	19
Non-structural measures	6
Promotion of resilience (acquisition of buildings, increasing elevation)	–
Support of ecosystems and activities around them	–
Promotion of compatibility with other actions	–

Table 4.2. *Restorative actions financed in the Coastal Master Plan dated 2017 [CPR 17]*

Coastal Louisiana is essential for the United States economy, for its history and for its culture. Some have envisaged reconstructing the only infrastructure far inland, at great cost and at the price of the "loss of communities, of the destruction of natural resources and of cultural dissolution". In other words:

> "If politicians fail to grasp the economic realities and citizens and thus engage in a long-term program of barrier islands, of coastal protection, the human landscape of the Louisiana coast will in the end no longer only be represented by a dispersed and nostalgic collection of memorial relics and memories of missed opportunities" [LAS 05].

4.7. Conclusion

The paradoxical title of this chapter, "The Aging Delta of a Country in the New World", was chosen to mean that the Mississippi Delta, whose capital was founded 300 years ago in 1717, is a young delta due to its history, but prematurely aged following development choices made for it.

The historical studies that were carried out on New Orleans point out that, certainly, Baton Rouge had the best site, but the location for a large city of the future was that of the current capital of Louisiana. Between site and location, location prevailed, without any real debate around the issue. The intrinsic fragility of the New Orleans site, built on a frail levee aggraded between the river and the Lake Pontchartrain, showed itself at the time of numerous river floods and rises in the level of water in the Gulf of Mexico during devastating hurricanes. The policy of development of the valley and the river, in the same way as that of the affluents, had a certain logic, in the reconciliation of trade and agriculture, but they have the insidious and progressive effect of reducing the capacity of New Orleans to resist high waters.

Economic development has had the effect of concentrating waters in the main Mississippi channel and in the Atchafalaya branch, consequently isolating the low-lying areas. The sedimentary fluxes, highly reduced by the reservoir-dams and by the deposition sites organized along the river, can no longer compensate for natural subsidence, aggravated by the extraction of hydrocarbons, even while the oceanic level rises; the areas of open water are spreading and the coastline is receding. Hurricane Katrina confirmed the weaknesses of the delta's capital and environment.

The future of the Mississippi Delta poses questions about the reality and sustainability of the low levels of sedimentary inputs coming from the watershed and methods to better manage the existing floods. The debate is sometimes lively, the solutions are innovative, if not always realistic, sometimes conflictual, pitting the reconstruction of marshes against a laissez-faire approach, the management of deltaic branches, the coastal defense or the organized recession of society. The Mississippi is the epitome of a laboratory delta, watched with interest by managers of the world's at-risk deltas.

5

What Strategies Can Help Overcome the Delta Crisis?

Above and beyond the shared observation proclaiming that the deltas are receding, sinking and losing their terrestrial attributes, in short that in the majority of cases they are under threat, their diversity defies comprehensive summary as soon as we go further from the case study stage. The future of the Earth's deltas appears relatively negative in light of the collapse of some of them, such as the Indus, and of the very worrying signals emitted for several mouths (e.g. the Nile and the Colorado). However, this vast image is not entirely bad, because deltas are still holding up well, and the increased awareness among scientists and managers, often genuine, is already leading to positive actions to slow down or inverse unfavorable trends.

5.1. Delta dynamics: contrasting budgets on a global scale

Over the course of the examples analyzed in this book, a series of change indicators have been presented and a deep complexity has been revealed, due to the great differences in geographical situations in the reality of "basin–delta couples", especially if we include, as we have tried to do, a certain level of historical background. Now let us try to draw up a summary.

5.1.1. *The progress of analytical approaches adds complexity to the understanding of deltas on a global scale*

5.1.1.1. *Basins and coasts*

No two deltas are strictly comparable in terms of natural conditions controlling their liquid and solid fluxes, in terms of the climatic history of their watershed recorded for more than 6000 years, and lastly in terms of their history of value being added to them by human societies. Complex interlinking of variables is a highly diverse factor and explains diverging contemporary trends that have a considerable impact on deltaic operation. But the overbearing trend does lean towards a reduction in sedimentary input at the mouths, even though they are the condition *sine qua non* of conservation of deltas as megaforms and as the basis for a rich economy and numerous human societies.

5.1.1.2. *Deltaic complexes*

Entering into the dynamics and the set of interactions at the heart of deltaic complexes, we see that they are controlled by individualized (or analytical) parameters that can quite easily be known in qualitative terms and are measurable with increasing success, even though on a local scale, reality sometimes prevents, or at least disturbs, scientific investigation. Let us remind ourselves of what the control factors brought together in the deltaic complex are.

At the scale of the fluvial systems that supply the deltas, there are upstream control parameters such as the flow rate entering the delta (the liquid flux is measured in terms of its volume, its fluctuations, as well as in the quality of the waters), the solid flux expressed in volume or in tonnage with the relative proportions of pebbles, sand and mud (made up mainly of silts and clay), or of particulate organic matter (the gigantic accumulations of tree trunks on the beaches of British Colombia in Canada, originating from the forests of coastal mountain ranges, are a good example of this). We are in the presence of factors that can be described as external, or control, because they are produced and somewhat imposed by the fluvial basin. It goes without saying that the time variable has a large role to play at the centennial scale (hydroclimatic variations condition the liquid, solid and organic fluxes) and at shorter time scales (wet years, high water episodes), but also anthropological pressures or their release into large watersheds; we have observed the great significance of the clearing phases in the Huang-He basin, in the European Alpes and the Apennines, and also more recently in the Mississippi basin.

5.1.1.3. *Oceanic control parameters*

The amplitude and twice-daily action of the tide has a powerful effect on the estuarian channels; similarly, the action of the swell and waves on the deltaic coastline when storms occur in the sea; the action of coastal drift; flows of fluvial sediments and sediments of marine origin in the channels and at the surface of the deltaic plain; saline intrusions into the surface waters and the underground waters, and so on. All these parameters interact at different paces, that of seasons and that of meteorological accidents that may be grouped together in periods or in a phase of high or low intensity; they themselves remodel the coastline and its deltaic inland areas.

5.1.1.4. *Internal factors within the delta*

Natural subsidence of the Holocene deposits depends on several parameters, among them the thickness and the lithostratigraphy of sediments; sometimes, the tectonics of the previous substratum and isostatic movements in the Northern hemisphere are added to this; the latter are more marked at higher latitudes, in places where the ice reduces its pressure on the Earth's crust and allows it to rise and cause differential movements at the edges. The environment of a delta is therefore by nature unstable.

5.1.1.5. *Direct human actions or anthropological forced influences*

Human actions and forced influences of various types have been playing a major role in many deltas, at least since the 19th Century. Extraction of water and of hydrocarbons trapped in the sediments lying at depth in the deltaic plain, extractions from exploitable sediment channels, construction of dykes and modifications of the distribution of flows at the surface of deltaic plains, alteration of plant formations in the mangroves of tropical deltas and water tapping for irrigation are examples that we have found in various places around the world.

The complex systems arising due to human interactions, of which each delta is a particular case, are part of the deltaic system. Less than two centuries of human history are all that has been needed to destabilize certain deltas, and their fragility has made this easy. In each of the examples studied, it has been possible to establish a chronology, predatory systems and internal geography, which have destabilized equilibriums that had been established with difficulty.

5.1.1.6. *The rise in ocean levels*

In itself, this constitutes a major impact on deltaic operation. Eustatic rise is itself the product of global climate change, where this is under the control of human actions through emission of greenhouse gases (to simplify), reasons for which we can talk about forced effects from anthropological impacts.

Each delta around the world is unique, because their set of external and internal parameters plays out their role there in a specific time frame, because the combinations of parameters in it are never identical and lastly because the relevant parameters for each one are expressed there with a unique intensity.

5.1.2. *The unforeseen effects of scientific choices*

The hierarchy among the factors varies spatially and temporally, as we have seen, but it can be disturbed by alterations. One of the aspects of the complexity inherent to fluvio-deltaic systems is indeed the competition that we observe between the explanatory factors as a function of the scientific specialties that are in fashion. The Mekong is an illustration of this point, which could well end up with non-negligible consequences in terms of management choices. Recent research that we have carried out for the entire Lower Mekong has highlighted the importance and the brutality of approximately 20 years of extraction of sand and gravel resources from the riverbed; we have also briefly discussed the underestimation of sand transport during the monsoon flood [BRA 13a, BRA 13b, KOE 12]. Lack of knowledge of this flux has largely facilitated exploitation of sand for decades: the river was (apparently) transporting so little that the act of extracting could not damage the equilibrium of the delta. Extraction of sand on a massive scale has provided an abundant and free resource; by digging out the fluvial bed, the extraction also presented the advantage of improving navigational conditions and of reducing the risk of flooding, arguments which are frequently encountered. On the other hand, revision of the figures for suspended and bedload transport has allowed us to better understand that the extractions interrupted the solid flux and have been directly responsible for the damage caused to the banks of the deltaic distributaries of the delta and of retreat of the sandy coastline.

Competition between the control variables for fluvial and deltaic operation has come to light through a modeling study that linked the intensity of erosion of the banks of the Mekong and certain climatic parameters; this is the snow melt and surface flow linked to the intensity of tropical cyclones, themselves under the control of El Nino/Southern Oscillation (ENSO) [DAR 13]. A more recent study by the same authors is even more explicit, insofar as its title directly links the reduction of sedimentary input into the delta to spatial fluctuations in cyclonic activity [DAR 16]. The conclusions of this study are that a third of the suspended load that arrives in the delta is due to cyclonic rain and that the load reduced by 52.6 ± 10.2 Mt between 1981 and 2005, and, for this figure, 33 ± 7.1 Mt are due to changes in the “climatology of tropical cyclones”. The study concedes that part of the load is trapped due to storage in reservoir-dams that were present in the basin in 2005, but that climatology does appear to play the key role. Other than the fact that the suspended transport data used in the study are rather debatable, because they are old and have not been revised, the role played by the extractions is not mentioned.

Thus, the climate is presented as the factor mainly responsible for the delta’s woes, which puts human responsibilities into a relative context of lesser impact, whether Chinese or Laotian upstream (the dams and blocking of the sand) or regional (extraction by neighboring countries downstream). Funding for construction works related to the effects of climate change is certainly of great scientific importance and may justifiably be sought by research teams, but it can also lead to side effects, since some could consider that financing delta restoration and modifying disputable practices (extraction) may therefore only be of secondary interest.

5.1.3. *Open, vulnerable systems*

The two most recent decades have allowed great advances to be made in terms of knowledge, but information in this area is still missing. In a first instance, still little is known about the real quantities of sediments that are deposited at the surface of the deltaic plains and those that reach the ocean, with notable exceptions; the figures are uncertain when the sediment dynamics are unstable: deltas become less effective as sediment traps as they undergo development and as the insidious impacts caused by the changes in the watershed modify the geometry of deltaic channels and in doing so the overflow conditions.

The sustainability of a delta is no longer guaranteed when its vertical construction by input from the basin can no longer keep up with the pace of the rise in sea levels. It now appears to be recognized that nearly all deltas larger than 10,000 km^2, and even deltas of an intermediate size with an area between 1,000 and 10,000 km^2, will benefit from continental inputs that are too low to be capable of resisting the effects of the rise in oceanic levels, even in the case of deltas whose fluvial system of tributaries has no reservoir-dams. In illustrative terms, these are "famished lands"[1] (to be understood as starved of sediments) [GIO 14].

A delta is a system that is especially open in nature and that requires fluxes of energy and matter (sediments, organisms and dissolved substances) to guarantee longevity of its correct operation and organization. The energy penetrates in "impulses" by means of floods, diffluences, construction of new lobes, breaches in the levees, through the twice-daily tides and through storms. The sustainability of a delta supposes that the forms of equilibrium created by the complexity of fluxes and of their effects are maintained. Thus, the Can Gio biosphere reserve, located on the eastern margin of the Mekong delta[2], created in 2000 by UNESCO, protects a mangrove of an area of 380 km^2 located in the region of estuaries to the south of Ho Chi Minh City (total area of the reserve: 720 km^2); this mangrove shelters a population of a density of less than 100 inhabitants/km^2, who carry out tasks of mangrove restoration and production of fish, shrimps and shells, in exchange for payment; the Can Gio mangrove produces goods and services which directly benefit the population and guarantee protection of the coast against the ocean, stores carbon, purifies water and soils, and finally constitutes a cultural heritage. The total value

1 By analogy with the expression of the clear water effect in French publications and of sediment starving, applied to rivers that are blocked and suffer from sedimentary deficit; they actively compensate for this through erosion of the bed and/or the banks of their channel. In the case of deltas, they are not able to adjust, at least not more than passively and by becoming exposed to fluvial overflows and to intrusions from the ocean.

2 This is the delta of the Dong Nai/Saigon river which adjoins the Mekong Delta, to which it is fused.

produced largely exceeds one billion dollars per year, precisely because the system is open and correctly operates despite the very high pressure put on it by the shrimp farms and by tourism [KUE 13]. The same potential probably exists for the Cape of Ca Mau.

Unless it receives various forms of energy of natural origin, a delta deteriorates and its inhabitants must use fossil fuels at increasing costs [DAY 16b] in order to maintain or develop their resources. A delta that has been entirely made artificial, such as the Nile, with disused branches, no flooding and which suffers from the effects of subsidence, with degraded soils and relatively low levels of marine intrusion, now urbanized and crossed by communication routes, is a delta in crisis. This delta no longer has the operational methods that are capable of providing it with a sufficient degree of ecological, and therefore economical, resilience, since the flow rate transiting through deltaic branches does not allow any erosion dynamics or any flooding to be expressed on the plain. Among other negative effects, there is a deterioration in the quality of the soils that can be observed in pedogenetic studies.

Above and beyond the shared observation according to which deltas are receding, sinking and losing their terrestrial attributes, in short that most of them are under threat, the diversity defies comprehensive summary as soon as we move away from the case study stage. How should we proceed from here?

5.2. Some control logic for rivers and deltas

In recent years, there has been a flurry of attempts at an overview, on the basis of combining homogeneous criteria.

5.2.1. *Situations involving crises and knowledge*

Analysis of some rivers and some historical routes has demonstrated that the understanding of upstream–downstream dominant relationships only come to light in situations of crisis. For this reason, it was of interest to analyze the crisis of the Little Ice Age in the first volume of this book, since it revealed the organic connections in place between the erosion of mountains and clogging of deltas by sediments. The understanding of engineers and foresters was based on several centuries of observations in the field and on refinement of knowledge that has progressed constantly until operational solutions are achieved. Recently, forestry practices from the 19th Century concerning erosion and reforestation have been severely criticized due to the fact that they participate in the current fashion of "eco-catastrophism"[3]. On the contrary, we have sought to demonstrate how a repetitive hydroclimatic crisis has led to considerable scientific progress, to the emergence of new concepts, and how this has produced remarkable solutions which have had long-lasting effects for the neighboring societies. It is possible that the most significant progress – the example of Renaissance Italy comes to mind – was based on naturalistic foundations rather than only on above-ground calculations.

5.2.2. *Contemporary hydraulic engineering pitted against the dynamics of economic domination*

The evaluation of the Mississippi and its delta has shown that the all-powerful nature of the US Corps of Engineers had had some undesirable effects, whether it was the case of the management of flood risk in the 19th Century or of Hurricane Katrina in New Orleans, of the distribution of flow rates between the main deltaic channel and the Atchafalaya, or even of drainage of the marshes to increase their agricultural productivity. Critics from the 1920s, and some more recently, agree that the errors of American engineering will have long-term effects, but have prepared people for the concepts of integrated management at the scale of the watershed and also for ecology, but more at the scale of the plains and the delta than the entire basin. We recall the phrase, already cited in Chapter 4 of this volume, in which the need for a "system that is integrated as such is insisted upon, and not as distinct elements that technology could manipulate" [KEL 03]. The trend today is to listen more to NGOs and researchers who advocate approaches that are much closer to nature and less costly than those of the USCE, upheld in the direct continuity of its cultural heritage. Similar remarks have been made on the subject of certain works in the Delta Plan in the Netherlands.

Issues that are of concern to developed countries are of even greater concern to developing countries, where engineering is often implemented without precautions. Previously, we mentioned some of the errors committed on the Yangtze and the Huang-He; they have caused considerable effects at great cost. However, this being said, China has developed its knowledge of the technology involved in large infrastructures; has it not invested even more in this area than in the delicate question of environmental issues? When we ask the question in this way, we already have part of the answer. The structures constructed on the Lancang – the upper

3 It is an oversimplification to summarize the question of erosion and of its treatment as a need for recognition expressed by a body of forestry engineers who are concerned about their power over submissive mountainous populations, nor is it credible to deny the reality of a crisis that is recognized and well documented in several fields. The promotion of certain debatable forms of catastrophism does not prevent the existence of true ecological disasters [VEY 10].

course of the Mekong – have had multiple objectives: exportation of Chinese electricity that has led to construction of an electricity network; integration of neighboring countries into the system of regional production, with Myanmar at the top of the list, followed by the Lower Mekong countries; or even construction of a navigable watercourse which is inserted into the infrastructure investment system of the BRI (Belt and Road Initiative, now OBOR for One Belt, One Road); it is very much the case of economic and political integration which has just won an important battle over the Mekong River Commission (MRC), automatically marginalized by the (more or less organized) division of the neighboring countries. The international question has been voluntarily placed in a regional context, namely simply the relations between the province of Yunnan and countries of the Lower Mekong, but there is more at stake, because it is really a case of systematic exportation of Chinese expertise and knowledge where the environment is not necessarily a sincere subject for concern. This is the mission that has been entrusted to the state company Sinohydro Ltd., the most powerful global company in hydraulic engineering, with the support of the Asian Development Bank [ALE 14].

In light of this, how much weight does the question of the fluvial environment carry and specifically that of sediment management of a river that supplies a delta? On the Lancang, dams have been constructed without sluice gates at depth that have the capacity to guarantee continuity of solid fluxes; it is true that the structures are so high that the lowering of the water level, required for a flushing operation, would suppose such a great loss of energy that it is hardly possible for an operator to entertain the idea. The Xayaburi project, prior to the expression of the Chinese policy that we have just outlined, is a Thai initiative dating back to 2007. The CH. Karnchang Public Company began work on the Mekong in Laos in 2012, in spite of great opposition from countries downstream and from environmentalist NGOs, and with disdain for the MRC's point of view; it was supported by European expertise, firstly from the Swiss company A.F. Calenco, then from the Finnish company Pöyry and lastly from the Compagnie Nationale du Rhône (CNR) for management of flushing, because sluice gates at depth were envisaged. The assessments carried out by the CNR justifiably highlight the role of the complex geomorphology of mountain watercourses in tropical regions and of the presence of sand that reduces the cohesion of deposits, in particular in reservoirs:

> "Maintaining a sufficient sand supply from the upper basin is of utmost importance for the long-term channel stability. The definition of a global management plan for sediments appears to be a relevant solution to achieve such a purpose… [but] in such a context, the sand inputs downstream of the Yunnan Dam Cascade are of prime importance for the sediment balance of the Lower Mekong River" [PET 14].

The CNR, which designed a hydroelectrical development plan for the Lower Mekong in 1994, thus came back into the picture, and has since joined up with Sinohydro Ltd which undertook the construction of the Sambor Dam, among others in the Mekong basin. We observe that Western firms are seeking to pick up pieces of the cake in Southeast Asia and invest knowledge in it to promote the most virtuous practices[4]; but their expertise, which will without doubt not allow correction of the multiple environmental impacts that are already in evidence, is obviously picked up by companies and governments who develop the installations on their territory while disregarding the rules of democracy. Western companies thus put their reputation on the line, export precious knowledge and will very quickly be copied.

5.2.3. *Scientific knowledge at the service of policies on rivers and on their deltas: the case of the Mekong*

Could the Mekong, which we have encountered several times in this book, not be fundamentally a victim of a threefold ignorance: one instigated by China, which does not communicate about the Lancang outside its borders; one due to the mediocre quality of the sediment measurements made at the beginning of the 1960s, measurements which were managed by the Mekong Commission after 1995 and which have persisted without too many inventories up until 2010; lastly, one from the massive sand extractions carried out in the river and which have benefitted suspect companies, badly thought-out projects and countries outside the basin with unscrupulous practices. On the back of this ignorance, projects for dam “cascades” on the river have been able to flourish; on the basis of disputable figures for sedimentary fluxes, scientists in developed countries have been able to draw up scenarios of the future amounts that all appear acceptable, and engineering has been able to put its skills to good use to overcome all difficulties and to back projects, some of which are bearers of the unknown.

It is perhaps too late to act in the Mekong basin, because the large dam projects have already begun. Even though the World Bank’s retreat from large hydroelectric projects at the beginning of the 21st Century provided hope of obtaining the time required to gain more in-depth knowledge, even though the government of Vietnam has firmly pushed for a moratorium that would have allowed consolidation of knowledge that is too partial, this has not been enough. Only the free riders Laos and Thailand, which freed themselves from the rules set out in 1995, have opened the path to China’s aggressive economic policy, which never recognized them and which

4 The Compagnie Nationale du Rhône has invested, along with the IRSTEA research institute, in knowledge of the transport of fine sediments in the Xayaburi reservoir to optimize the flushes that have been made possible by installation of sluice gates at depth.

launched its OBOR (2013) initiative, then the Lancang–Mekong Cooperation Mechanism (2016). China's hydroelectricity-friendly energy policy is legitimized by the recommendations of international conferences on climate change which are in favor of the reduction in greenhouse gas emissions, a policy that lifted most inhibitions, but without completely and critically appraising the negative consequences of energy from hydraulic sources in environmental terms. A key point is sediment trapping, as this book has attempted to explain, and everything possible should be done either to minimize its impacts or to avoid creating them, simply by protecting the watercourses. Environmental regulations should be toughened to progress towards resolution of key questions for the future of rivers and deltas.

Inversely, it is the case that the most advanced scientific knowledge does not provide much assistance by way of proactive management. A financial study by the US Agency for International Development (USAID) has modeled the cumulative storage of fluvial sediments in reservoirs under construction, at the scale of the Lower Mekong and of its tributaries, and in those that are planned. Under the hypothesis of constructing 38 dams (mainly placed on tributaries), storage would be 51% of the present sediment flux at the mouth; under the hypothesis of construction of all of the 133 planned structures, the cumulative storage would be 96%! The logical consequence of this catastrophic scenario becomes not the selection of the least dangerous structures but rather the quest for solutions to reduce this entrapment [KON 14]. The following step, carried out as part of the Natural Heritage Institute financed by USAID, consisted of modeling, over a period of 100 years, entrapment in reservoirs along three tributaries (the "three Ss") that descend from the Annamite mountain range (Se San, Sre Pok and Se Kong). No matter what the method of management for the structures, entrapment remains low, does not increase further after 100 years and remains at approximately 20% of the load if sluice gates allow flushing. In the end, the conclusions are in favor of the construction of structures equipped with sluice gates, because they would have relatively little influence on trapping and energy production if correctly managed. But research goes even further:

> "Future work must include better understanding the interactions among sediment, nutrients, and aquatic species in the Mekong basin. Until these interactions are better understood, it is difficult to estimate the true economic and ecological benefit of sediment management" [WIL 14].

In other words, nothing yet appears to prove that a lack of sediment management is a negative thing; according to the authors, future spending to improve the life of species will only make sense if it can be redeemed; otherwise, sediment management will be useless. Here, we can observe all the dangers of a technical and technocratic approach from which the delta, as a highly complex environment, is also absent.

5.2.4. *Avatars and tribulations of geopolitics*

One of the most important aspects of river management, and indirectly of delta management, is the effect of upstream dominance, whether it comes from several states or regions or from one single state. This is a well-known aspect that arises from "hydropolitics" and now from "hydrodiplomacy", encouraged in particular by the SDC, the Swiss Agency for Development and Cooperation; we will not explore the subject further in the restricted context of this book, although it has been touched on several times in the previous chapters. Over the course of the book, we have encountered the effect of dominance on hydrosystems such as the Colorado (United States and Mexico); the Ganges and the Brahmaputra (China, India and Bangladesh), of course the Mekong, the Rhine (Switzerland, France, Germany and the Netherlands), etc. We also cite, without developing it further, the case of the Tigrus and the Euphrates, controlled above all by Turkey, to the detriment of the delta of both rivers, in a geopolitical context that is too complicated to be presented here. On another river, the Senegal, President Macron had the threats presented to him in February 2018 in the town of Saint-Louis, in particular the breach of the Langue de Barbarie spit that protects it, but only the rise in sea levels caused by climate change was presented to him[5], not the modification of the hydrological regime and sediment storage by the Manantali and Diama dams constructed on the river. It is true that this simplistic view, because it is partial, has the merit of being effective; it has allowed funds to be released to protect the town of Saint-Louis.

The Nile is very particular, in that the dominant country since the 19th Century has been Egypt; supported by treaties drawn up and encouraged by the British, Egypt has imposed its military strength on the vague hopes of development entertained by upstream countries, at least until the intervention of Chinese investors and the construction of the Great Dam of the Ethiopian Renaissance, the most powerful in Africa at 6,450 MW with a height of 175 m, and with its reservoir capacity of 67 km^3, on the Blue Nile by an Italian company (2013–2018). The Egyptian disagreement is demonstrated by reliance on historical treaties, but with no ability to block the project supported by the upstream countries founded in the Nile Basin Initiative. Since the Ethiopians have refused all impact studies, international backers have refused the funding, but have been relayed by China. A unique situation, of

5 M. Meybeck, *in litteris*.

complete dominance by a downstream state, is in the process of being reversed, but nothing is yet set in stone.

Within a single country, it is rarer for the situation to be analyzed but, other than the competition between the countries in the Colorado basin in the United States, one European case of upstream to downstream dominance can be mentioned. This is the Ebro basin, where the delta has lost more than 800 hm^3 to Aragonese irrigation and basins in the south of Spain (Jucar, Segura and the region of Almeria) through application of the 2001 National Hydrological Plan.

5.2.5. *Expert appraisal and conquest of engineering markets on deltaic land*

In recent years, there have been diplomatic innovations that have added the weight of scientific groups to the approach adopted by the Ministry of Foreign Affairs. "Scientific diplomacy" was born in London in 2009, with a view to international agreements, for example, in environmental terms [ROY 10]. In fact, climate change has mobilized a high-level international expertise since 1998, under the framework of the IPCC, at the service of multilateral political decisions. But:

> "In international negotiations about climate change, it is not science that has the last word, but national interests, and it is difficult for them to become part of strong collective commitments: science plays only an imperfect role in climate diplomacy. And it is also national interest that can lead a country to rely on climate research to improve their own diplomatic positions, thus illustrating the supporting role played by science in diplomacy" [RUF 18].

Yet, in recent years, we have gone from decision-making based on scientific knowledge to finalized research, carried out to meet the requirements on the large political and private ordering institutions.

We would like, following on from the previous idea, to now broach a delicate subject, namely the alliance of diplomacy, science and technology with well-understood national interests. A question of global proportions, concerning deltas under attack from bad management practices in watersheds, is also reduced to a level of national concepts and benefits.

The skills and, in a more general manner, the expertise come at a high price. Countries and companies have understood, for a long time, that applied science is a very good export product. Modeling complex dynamics, designing management plans, organizing agreements that bring together the users or the stakeholders, implementing projects when they require technologies that are out of reach of the financial means of the countries concerned or when they are eligible for subsidies from large international institutions, and so on, all this requires faultless organization, mastery of communication to reassure, convince and carry off. In a single word, most of the time, it is a *business* based on high added value[6].

Other than international funds, such as those of the World Bank, destined to help emerging countries, there is very little money for international research and each country generally pays for the contribution made by their researchers. Large companies benefit from the subsidies by presenting themselves as the most competent to carry out the works. However, on a national scale, companies have quickly understood that if they invest sums of money into "research and development", the countries multiply these amounts in various ways, like doubling the amount put into public–private partnership programs, tax credits or simply by letting companies use personnel and public services (e.g. laboratories) without paying for the complete costs of research[7].

Domination of a market means, as in the example of the Netherlands, being able to associate complementary strengths in an effective manner to use in joint projects involving water and surficial geology (soils, subsoils and water tables). The highest rung of the conquest system is the government that delegates specific missions to the Ambassadors: analysis of the situation on site (risk, demand for economic development, environmental imbalances, etc.), integration of the Netherlands into international structures, the attribution of subsidies or funding of assessments and key-in-hand pre-projects. Once the contacts have taken root, the principle consists of allocating study grants and training future engineers and technicians in specialized universities in the Netherlands (e.g. Leyden, a large training center in hydraulics) or through international lessons given abroad; uniting university strengths (training, research); putting universities in contact with design consultancies and engineering companies that carry out the projects and communicate in the requesting country;

6 France proceeds through the intermediary of the French Development Agency (AFD), set up in 1998 to take over from the French Development Fund (CFD). This is a public financial institution that acts to reduce poverty and encourage development. The AFD has a financial branch, PROPARCO (Promotion and Participation for Economic Cooperation), that brings together private and public sources of funding. The money loaned is supposed to support projects that are "economically viable, socially fair, sustainable from an environmental point of view and financially viable". In addition, the International Office for Water (OIE) hosts networks of actors and manages many projects around the world.
7 Yves Bégin, *in litteris*.

and lastly, incorporating Dutch structural design laboratories in this organization (small structures), then companies that are capable of carrying out large construction projects.

We are in the presence of a vertical organization financed by private companies or by the Dutch government under the framework of "strategic" projects; within a few years, this organization emerged from the national territory to become a true empire. Management is based within the institute of applied research DELTARES, whose slogan is *Enabling Delta Life*. Based in Delft and Utrecht, this organization of 800 people, which is described as independent, has a far wider target than their slogan appears to indicate: water resources, management of fluvial basins, deltas and coastal regions, infrastructures and the environment, and water architecture; in short, it maintains its vocation to work in almost all environments around the world. This public institution is the iron fist of Dutch scientific and trade policies. It is relayed by powerful anonymous companies like ARCADIS (Amsterdam, 27,000 people all around the world), Royal HaskoningDHV (Amersfoort, 6,000 people) or even BoskalisNV (Papendrecht, 11,000 people); these companies undergo rapid external growth by buying competing companies, in particular in the United States. Architecture design and urban planning consultants participate in the consortium, like the company Kuyper Compagnons (Rotterdam, more than 100 employees), which succeeded with several calls to tender for innovative projects on aquatic sites in China, Singapore and with the Garuda project in Jakarta (see below).

Dutch NGOs like SIMAVI, IRC, AKVO and Wetlands International are reasonably in favor of measures known as "non-infrastructural", therefore softer. In 2016, in Stockholm and during World Water Week, they launched a "strategic partnership" with the Dutch Ministry for Foreign Affairs, entitled End Water Poverty, an interesting approach combining both generosity and potential profitability in the long term for the Netherlands as a company[8]. The complexity of the approaches is further reinforced by the action of research institutes, such as UNESCO-IHE based in Delft; created in 1957, it is the only institute financed by UNESCO that is accredited to issue master's degrees and PhDs. The largest institute in the world specializing in the field of water, it trains young researchers and managers of developing countries in integrated management.

All Dutch water management organizations are united in a group known as Dutch Water Design. The Netherlands thus became a "nationwide consulting

8 WASH Alliance International ("wash" is a contraction of watershed) aims to establish skills in civil society so that it can put pressure on governments and guarantee basic services related to water: from the supply of drinking water to the release of wastewater.

company" endowed with "water diplomacy" and considerably benefitting[9] from recent disasters and the rise in dangers related to climate change [GEN 11]. The declared principle of the Netherlands' "niche", certainly generous in its expression, is to contribute to the "development of global public goods". The Netherlands considers water diplomacy to be a priority for the country: they are already present with the Permanent Court of Arbitration and the International Court of Justice, which have had to treat the Indus conflict and the controversy surrounding the Nagymaros dam between Slovakia and Hungary; their Ministry for Foreign Affairs intends to take the lead in European and then global water management, in partnership with large financial institutions, the United Nations and the FAO.

Domination by the Netherlands is lastly based on an effective storytelling process in which the 1953 catastrophe is presented as the founding element of a collective behavior based in solidarity [KEE 16] and on a sense of resilience built up over the centuries; this dominance was increasingly reinforced by engineers, scholars and designers. The companies are much older than the mythical date of foundation of 1953, because they were born in the polders, but the sales pitch, recited often, is very persuasive. The Netherlands thus became the top engineering laboratory in the world and turned their expertise into their main source of income. They have been able to go from resistance to storms, based on powerful concrete structures, to a resilience based more on a compromise with nature: in the marshes of the Mississippi delta, in Jakarta in the same way as Bangkok, in New York affected by Hurricane Sandy, in the Indus, and so on. This bulimia is certainly able to establish itself, but it presents the considerable risk of suffocating knowledge, local initiatives (when not local companies themselves) in favor of a homogenized and formatted view.

Another risk, and not the least, is that in their quest for engineering markets, research-engineering consortiums attach excessive importance to applied research of a sectorial nature and intended *in fine* for specialist companies. Seductive solutions if we consider that the decision makers want to communicate in simple terms about projects that they can put in place in the short term and that the proposed technical solutions are subsidized by developmental aid. Interesting solutions above all for the DELTARES consortium, which simultaneously sells them key-in-hand in several

9 A "fantastically lucrative" system according to a partner of the setup [CHU 13]. Additionally, during regrouping of the head offices of the company Unilever in Rotterdam, P. Klok, writing for the center-left daily newspaper *De Volksrant* (16 March 2018), congratulated himself on this decision, flattering for the Netherlands; "but the best thing about the story, is that the Netherlands could help Unilever implement its sustainability strategy… by stimulating cooperation with universities and companies", he added. Another example of a strategy that in all directions aims to reinforce synergies between Dutch partners.

countries around the world. With this reductive approach to the complexity, undesirable side effects can occur: as a result, managers and politicians could focus on single factor, or at best partial, causalities, to the detriment of the basin-delta complexity. How about, for example, the sedimentary deficit that affects parts of the Mekong Delta, which are sensitive to it, if a project like Rise and Fall, related to the effect of pumping on the future of cities and on drinking water downstream from the delta, focuses too much attention on the subsidence induced by overpumping and not on the system of factors that form the underlying cause? Here, we again come across, along with other factors, the question asked in Volume 1 of this book (Chapter 4), concerning the dangerous level of importance attributed only to the change in cyclonic conditions, to the detriment of the enormous extractions of sand from the bed of the Mekong, whose negative effect appears to us to be at least as significant. Once again, the quest for engineering tenders threatens to prevail over an understanding of the complexity of the affected natural systems and the quest for truly sustainable solutions.

5.3. What sustainability is there for deltas in the 21st Century? Comparative approaches

What can the sustainability of the Earth's deltas in crisis be, knowing that the main challenge that they experience is "human occupation in a transitory environment"? [DAY 16b]. This question is asked by several research centers who are seeking to overcome the complexity of the analytical approaches presented at the beginning of this chapter.

5.3.1. *The typology of deltas as a function of the changes expected in the risk profile*

An overall (and quite sophisticated) approach was proposed with the aim of leading to a quantitative and comparative evaluation of the risk to deltaic companies around the world, while at the same time, this evaluation intends to take into account the change in natural and anthropological forces in the future [TES 15]. An overall indicator is opposed to two values, the vulnerability and the exposure, which when combined give the predictable variation of the risk to which the populations could be subject.

The exposure of people increases with two components of the hazard, namely the relative rise in sea levels and any worsening of flooding; the first depends on an indirect anthropological control (e.g. subsidence of the delta due to an impact effect). The future vulnerability will also be a function of the amount of GDP or international subsidies that the country will be able to activate in order to influence the implementation of the measures to reduce the vulnerability, and the governance of the territory; it is expressed by an index that estimates the probable investment deficit with regard to the increase in the 1 in 100 years hazard (Table 5.1)[10].

Factors increasing the risk (R')	**Deltas affected at 100 years**
Average increase in vulnerability (V), low increase in exposure (E)	Mackenzie, Lena, Yukon
Increased vulnerability (V) and exposure (E)	
– Very large increase in V and E	Rhine, Mississippi, Yangtze
– Large increase in V and E	Rhône, Parana
– Average increase in V and E	Amazon
Increased exposure	
– Average increase in V, large in E	Ganges–Brahmaputra, Krishna, Indus
– Average increase in V, average in E	Mekong
– Low increase in V, average in E	Magdalena
	Limpopo, Fly

Table 5.1. *Classification of the Earth's deltas as a function of the criteria for predictable increase in their vulnerability and exposure of people [after TES 15, modified]*

5.3.2. *Typology of deltas as a function of their energy consumption*

Another typology was proposed in order to classify deltas according to the two large challenges of the future, namely demographic and economic growth on the one hand and climate change on the other hand. The criteria for the typology are the climate of the fluvial basin (arid, wet tropical); the resilience of life in the country and its relationship with the waters to be combated; and lastly, at the scale of the delta, the increase in human impact and the importance of the

10 The risk (R) is defined thus: the hazard (A) × exposure (E) × vulnerability (V). The hazard is the probability that a damaging event will occur; the exposure is the number of people exposed to a hazard; the vulnerability is the damage or the loss caused by the hazard.

surface area below the level of the sea [DAY 16b]. This typology is difficult to select, because it has the disadvantage of being based on interlinked criteria that are heterogeneous from one type of delta to the next and which are unsatisfactory as a result. However, we will select the distinction made between the deltas for which management does or does not consume energy and their capacity to ensure sustainable energy management in the future. This is a criterion that is indeed important.

Deltas with management that consumes large amounts of energy

Rich countries invest considerable amounts of money in protection of their deltaic space against fluvial flooding and/or against the rise in sea levels and storms; for example, the deltas of the Rhine and the Meuse, the Po, the Mississippi and the Yangtze, where expenditure is allocated to dykes, sluice gates and pumps. The hypothesis is made that the increase in natural constraints will require increased expenditure in the future, to such an extent that the sustainability can no longer be carried out everywhere. However, we note that this pessimistic hypothesis does not take into account a decrease in the cost of energy by resorting to wind turbine and solar technologies, renewable energies whose use would reduce transport costs.

Deltas with management that consumes low levels of energy

The deltas of the Ebro and the Rhône are the first to be affected by the innovative and economic scenarios whose implementation would be a break from current practices; the Danube and the Yellow River are also concerned. The suggested solution, which will be presented later, is based on the principle of reconnection of fluvial branches to paddies in the deltaic plain, with the aim of combatting subsidence by introducing fluvial sediments. These projects are seductive on paper, but appear a little utopic in the current state of things.

5.3.3. *The degree of vulnerability or the relative vulnerability of deltas to current changes*

A typology that proposes to group together a certain number of deltas into four classes of sustainability on the basis of knowledge and expert judgements is interesting. The criteria are the current state of the deltas, the intensity of impacts that they have already been subject to and their capacity to overcome the constraints of climate change and the growing rarity of energy (which, let us recall, should only be a hypothesis). The authors propose to develop a suitable metric in order to move away from considerations that are still qualitative (Table 5.2).

Degree of sustainability	Major criteria	Selected deltas
1) Deltas with high potential	Low sedimentary and hydric deficits	Amazon, Orinoco, Grijalva–Usumacinta, Parana, Irrawaddy
2) Deltas with moderate potential	Potential to reduce the constraints if offensive management and low-energy consumption	Yukon, Lena, Rhône, Danube
3) Deltas with low potential	High impacts on human activities, management programs that consume high levels of energy, extended sectors below sea level, but persistence of fluvial deposits	Mekong, Yangtze, Ganges–Brahmaputra, Mackenzie, Mississippi, Magdalena, Ebro, Rhine, Po, Volga
4) Non-sustainable deltas	Arid regions, reduced and reducing water and sediment inputs. Predictable worsening due to increasing impacts and climate change. Degraded or zero operation	Tigris–Euphrates, Nile, Colorado
5) Deltas that have suffered a collapse	Low and temporary fluvial flow rate, contraction of the delta, pollution and saline intrusion, depopulated villages	Indus

Table 5.2. *Estimation of the degree of sustainability that can be envisaged for some of the Earth's deltas (source: [DAY 16b], modified, the Irrawaddy has moved up from class 2 to class 1). Note: we have taken out the Congo Delta (which is underwater) and retrograded the Volga Delta, highly impacted, from category 1 to category 3*

5.3.4. *The notion of the tipping point of a delta and of the socioeconomic system*

Finally, a recent study has proposed to consider the Earth's deltas as socioecological systems that are subject to human interventions, more or less intense in nature and combined due to their attractiveness. This latter aspect, firstly agricultural and for fishing, now concerns multiple forms of development which meet in the deltaic plains and impose their impacts on it. The authors distinguish four states on an increasing scale of intervention: the Holocene state (geological or natural), the modified Holocene state, the Anthropocene (dominant human control state leading to complete alteration) and the state of collapse [REN 13].

The question is whether a system can absorb constraints or recover (through resilience); it is also when and where the system reaches a tipping point or a threshold beyond which irreversible processes occur (natural or anthropological) leading to its collapse. The "tipping point" can also be at a specific moment or at specific places in the delta, when the state of degradation of the system is such that a company admits abandoning the natural forces and active collapse, because it is then no longer capable of taking back control of the forces.

5.4. Actions at the scale of the continental fluvial system to rebalance the deltaic systems

The general principles of new environmental management are to restore the operation of the deltaic system by using ecological engineering techniques. It is a case of working alongside nature, in other words, to reintegrate natural impulses into delta management. In other words, to restore environments while minimizing fossil fuel inputs in favor of natural energy and to avoid disturbances of anthropological origin. These principles should have the advantage of providing a better income to local populations, whereas the beneficiaries of large construction projects are (often) political "elites" of the countries that are affected in situations of crisis and companies, often foreign [DAY 97].

At the scale of fluvial basins, the main recommendation is to maintain a sustainable export of sediments from the continent towards the deltas in order to restore sediment budgets that are compatible with the rise in sea levels. At a minimum, it is a case of stopping their deterioration. Maintaining natural formation processes and longitudinal transfer of flows occurs by means of voluntary actions based on the following elements.

5.4.1. *Implementation of actions of sedimentary management*

Correct sedimentary management identifies the sources of materials and fluvial sections that are likely to produce an additional load due to erosion of the deposits stored in the plains.

Above all, production areas must be closely considered in order to evaluate their capacity to supply significant transport (e.g. sectors of erosion in mountainous massifs and hilly regions). In the volumes produced by erosion, the complete granulometric spectrum must be taken into account, including sands, of which we know the importance for correct resistance of coastlines against marine constraints, silts and clays, for their importance in maintaining fertility of alluvial plains and deltaic plains, and lastly silts and clays again, for survival of the mangroves. In the Southern Alps, in

France, the Durance basin has thus provided considerable quantities of pebbles and silts over recent centuries in direct relation with the lithological nature of the basin (predominantly limestones and marls). Organized and spontaneous reforestation (related to the depopulation of mountains) has reduced the erosion, but not to the point of eliminating all the silt input coming from the ravined marls. Protection of reservoir-dams, waterways constructed for irrigation and hydroelectric production, and the Etang de Berre (the artificial exit for waterways from the Saint-Chamas power station) with regard to the deposition of silts, led to the pursuit of a policy of erosion control which was systematic at the scale of the basin, without fully succeeding in spite of ongoing efforts. This policy's logic lay in the prospect of sustainable economic management of hydraulic structures of the basin, but the progressive return of waters in the lower Durance (a positive feature) underlines the deficit of sediment fluxes that then reach the Rhône, the delta and the Mediterranean.

There is then the question of transport of particles in the branches of the fluvial system, from the sedimentary sources to the sea; according to the terminology of the day, this is the principle of maintaining or restoring "sedimentary continuity", knowing that it is not a continuous and homogeneous process, but rather a disjointed process, pulsed during the floods, consisting of deposition in the plains and inputs through lateral erosion of rivers when the latter still have the capacity to do this. This is because rivers in Western Europe have lacked load as a result of the changes in watersheds and became deeper along their courses in the 1950s–1980s (when their load and their bottom sediments have not been overexploited by the extractors) such that a movement associating researchers and managers has designed and locally implemented the notion of fluvial free space in which erosion was favored to re-supply the rivers (a little); this method has been effective to a certain extent to ensure clearance of deposits from historical erosive crises that have contributed to aggrading fluvial beds and the adjacent plains. Used for the first time on the River Allier, the largest tributary of the Loire, at the beginning of the 1980s, this effort to restitute mobility of river reaches was used on the Ain, a tributary of the Rhône, at the beginning of the 1990s; it is now implemented on the Rhône downstream from Lyon, where deposition of lateral dykes from the 19th Century allowed pebbles and sands to be picked up from the banks in certain old Rhône branches by-passed by the diversion canal, but always active in flood; it is not a case of "liberty" as such, but of a "reactivation of the margins of the Rhône" that liberates materials. It goes without saying that sedimentary mobilization cannot be designed without coordination between upstream and downstream, capable of managing all aspects of sedimentary continuity. In France, it can only be carried out by public management authorities, such as water agencies, or by concessionary companies in the public domain, like the Compagnie Nationale du Rhône, but even this scale is not enough, and participation from EDF (French national electricity service provider), which manages hydroelectric production on tributaries in the European Alps and in the Massif Central in France, is required in certain cases.

Restoration of the sedimentary fluxes disturbed by the dams has resulted in some structures being fitted with sluice gates at bed level to allow the materials stored in the reservoirs to pass through, but this type of management is complicated and not without ecological impacts downstream; the most notable examples are on the Yellow River, as we mentioned previously. In Europe, the case of the Rhône downstream from Geneva is of interest. Flushing of the reservoir-dams in the upper Rhône mainly has the objective of maintaining the hydroelectric production capacity of existing structures, in particular the Verbois Dam in Switzerland, then the Génissiat Dam and the diversion structures constructed up to 100 km downstream, but no further than this, in the direction of the sea. Purging does not take place during high water levels (the managers need to have quite high flow rates passing through a reservoir, which must have a reduced level, in order to give the flow some speed) but instead, in general, every three years at a suitable date, taking the hydrology into account, with controlled help from Lake Leman which provides the equivalent of a flow rate of very clear high water levels; in the reservoirs along the French section of the river, dredging assists the three-year clearing process.

A plan was created in the middle of the 1990s to partially correct the excessive retention of sediments in the hydroelectric structures in the lower Ebro valley (Mequinença, Ribarroja and Flix) and resupply the delta based on the Mississippi model. The average subsidence of the delta today is 2–3 mm/year, to which the relative rise in sea levels is added, which is 3 mm/year. This dynamic should lead to a relative rise in sea levels of 50 cm before 2100, which would flood 45% of the currently emerged delta. To counteract this evolution, which would require an aggradation of 1 cm/year, there would need to be 1.3 Mt/year, which is 10 times more than the current load, but 20 times less than the load before the dams [ROV 07]. But how can managers proceed? For example, during high waters, dredging of large sediments would take place in the upper part of the reservoirs with deposition of these behind the dam and close to the sluice gates so that they are evacuated towards the delta's coastline; during high waters, some of the finest sediments could be dredged, remobilized and sent towards irrigated rice paddies and marshes along existing channels. These structures were considered close enough to the sea for this to be possible, and a preferred idea was to dedicate a proportion of the water in the Ebro basin to ecological purposes, while Spain's Plan Hidrologico Nacional only made provision for economic uses that were believed to be a priority [IBA 97]. Flushing was attempted from 2002 onwards, after alteration of the liquid and solid fluxes led to development of macrophytes* on a massive scale in the downstream bed of the Ebro, downstream of the dams and over a distance of 85 km. Lessons have been very carefully learnt from this, and the procedure has not yet been carried out, because the transfer of materials takes place in a very complex manner, and it is a good idea to make sure the sedimentary "wave" is capable of descending downstream and to also evacuate the accumulations that are constructed at the confluence of the Ebro's small tributaries [TEN 12, TEN 14].

A prospect that seems appealing is restoration of the sedimentary and ecological continuity of rivers, which in Europe relies on the Water Framework Directive 2000. In France, the principle applied was to consider that river weirs brought the same disadvantages as dams. While the principle of restoration of the main migration routes had been laid down and was acceptable for certain watercourses, if not everywhere, some aspects that posed a problem remained: without an inventory, is it possible to break up a heritage that is often ancient, with parts dating back to the Middle Ages? Is it possible to choose a policy that denies the social demands in terms of heritage or leisure activities? Do we really re-establish sedimentary transport and worthwhile fish mobility by removing weirs? Are the released sediments going to reach the sea or are they deposited on the way? We can understand the interest of removing certain obsolete hydraulic structures, but a generalization to an entire country can call into question whether a systematic policy is well founded. And we still need to ensure scientific monitoring of the benefits of these highly costly operations. The principle of protecting the integrity of rivers that are exempt from development by putting them into the category of "wild rivers" poses less of a problem and leads to increased awareness among the inhabitants of major natural heritage, without, however, destroying a cultural heritage that is highly important for many towns and villages.

5.4.2. *Establishment of current and future sediment budgets*

Sediment budgets are essential, or should be essential, before large development projects are implemented. No serious official communications were carried out on hydroelectric development of the Yangtze prior to construction; however, recent Chinese scientific works, some before completion of the Three Gorges Dam in 2009, do not at all hide the impacts on the watercourse, most of which were predictable (Volume 1, Chapter 5). Whether for the Yellow River or the Yangtze River, Chinese published research and its applications are progressing in a remarkable manner.

The principle of carrying out prospective sediment budgets presumes that knowledge of rivers producing sandy and silty flows will increase and that they will be conserved, in other words, that a choice is made to not develop certain routes that are essential for transport of sediments towards the deltas because the techniques have not been perfected, or even, even though they are, because in certain countries there is no guarantee of correct management of structures for optimal transport. The future for deltas in basins that are undergoing hydroelectric development requires a guarantee of sustainable energy development.

How should we proceed? A simple idea is that it would be a good idea to plan the sustainability of natural sedimentary production processes in basins with positive sediment balances, which was, for example, the case for the Mekong before the beginning of the current wave of construction. A study highlighted the need for a sediment budget at the scale of the hydrological network of the Mekong basin, but the Mekong River Commission was not able to follow up on this; this is not the only study, but it is very late[11]. The question is to specify in advance what the socially acceptable erosive processes are. The materials in the basin that are useful for a delta are fine materials which partially come from natural erosive processes located outside agricultural environments; another part comes from agricultural erosion on steep, fragile slopes whose strength has often been weakened by slash-and-burn farming. It is considered legitimate to protect agricultural soils that are the backdrop to the life of rural communities. It is possible to envisage success of a policy of this kind, as was demonstrated in numerous regions of the globe, but to the detriment of the load and eventually of the deltas. It goes without saying that a secondary watershed, supplying a river and with little impact from hydraulic structures, can play a major role in sediment balance at the scale of the river and its mouth. But then, a detailed budget must be carried out in order for the components of erosion to be known relatively precisely before implementation of any policy at the scale of this basin. Input of materials, if it must continue, cannot occur in a manner that is detrimental to the resources of the rural environment.

An initial study financed by the World Bank is in progress on the Ayeyarwady in Myanmar, a river which has still managed to escape any significant dams despite Chinese attempts (this is not the case with some of its tributaries, however). It has a "Sediment and Geomorphology" section with an inventory that demonstrates the critical nature of the question, the weakness of the available data and, again, the need to invest in gathering knowledge before any decisions are made[12].

5.5. The actions developed in the deltaic system in response to crisis situations

5.5.1. *Structural solutions: dykes and fluvial levees*

5.5.1.1. *Protection of land*

It is rare that the fluvial branches of a delta or of its coastline do not have structures that partially or totally protect the deltaic plain from floods or storms. Dykes or fluvial levees have been described for the Mississippi where, in the same

11 A study [BRA 13b], financed by the French Facility for Global Environment, constituted part of the project carried out by the Information and Knowledge Management Program of the secretary of the Mekong River Commission and by the Greater Mekong Program of the World Wide Fund for Nature (WWF).

12 The sedimentary section of the Ayeyarwady State of the Basin Assessment (SOBA) was presented in October 2017 by WWF Greater Mekong to the Burmese Ministry for Transports and Communications.

way as on the Loire in Orleans, New Orleans built protection structures; we encountered them on the lower course of the Rhine and the Meuse, on the Danube, on the Mekong and the Irrawaddy. This is not the case for all deltas, because the fluvial branches have kept a certain mobility on the Brahmaputra, the Niger and Danube deltas, for example. The dykes are a characteristic trait of old fluvial societies in Europe and North America, and the technical affiliations are obvious. We see their unusual forms along the Lower Mekong, downstream from Phnom Penh, where the infilling from dredging raises and widens the natural levees and extends them laterally to cover vast areas of urbanized land or land devoted to industry. We can also cite the dyked agricultural areas in the Ganges–Brahmaputra Delta whose negative impacts and the absence of sustainability are disputed (Chapter 3).

5.5.1.2. *The sea dykes*

More recent than the fluvial dykes, they are quite often constructed against contemporary threats from the ocean or from the sea. A very old system of dykes is found in the *terra firma* behind Venice; the rise in sea levels and the breaches opened by storms in the barrier beach then motivated construction of powerful protections (MOSE project). This is the case for the Camargue dyke, constructed in 1859 as an addition to the fluvial dykes that were reconstructed after the great flood of the Rhône in 1856, or sections of dykes of the Mississippi Delta that protect sandy bars that have been damaged by marine erosion since the sedimentary deficit has appeared. Without backtracking on the Zeeland sea dyke model in the Netherlands, which replaced old sections of dyke after the 1953 catastrophe, the great novelty is the barriers on fluvial exit channels guarding against high tides and storms. This change of scale can be alluring due to the security it provides to the Netherlands. This concept was re-used in Shanghai against the Yangtze and the China Sea, and also in Jakarta.

The Garuda project (Great Garuda Sea Wall or GGSW), on an altogether different scale, has aimed since 2014 to encompass the constructions in progress along the banks of the Jakarta [NCI 14]. For a cost of 34 billion euros, construction of a dyke that is 32 km long and 24 m high is planned for 10–15 years to close the bay. The dyke will isolate a "lagoon" of 1,000 ha, in which the salt water will be replaced by freshwater and whose level will be maintained by pumps. Seen from an airplane, the general design of infrastructures will look like a *garuda*, a bird from Hindu mythology that symbolizes Indonesian royalty [MEZ 16]. The sea dyke will support the head and the wings of the bird (the head will be the business center at the center of a water front extended on each side by districts containing urban parks known as *wing parks*). The stretch of water will feature islands in a fan shape (the bird's tail) and a series of islands arranged in front of the current polders, based on the model from Singapore. At a cost of 40 billion dollars, the entire development will house 2 million residents and 600,000 employees, a field day for the property sector. The masterplan of the National Capital Integrated Coastal Development Program (NCICD), a plan that enjoys a high level of visibility, and whose design was financed by the Dutch government, groups together a consortium of design offices in the Netherlands and some Indonesian expertise[13].

This plan led to many critics that anticipate considerable social costs for the populations living in a traditional way on the coast (in particular, fishermen). The project would in fact be a giant operation destined to make the town into a "world-class" capital, based on power relationships at the highest level between the two countries, for a flow of private capital seeking investment opportunities, and on the action of a vast technopolitical network, while parties in the field of water are seeking new markets; the latter is in part founded on colonial and post colonial relations based on Dutch expertise and the development of knowledge, as technical as it is alternative, in the former Indonesian colony. The Garuda Giant Sea Wall (GGSW) project does not aim to fight against subsidence and to combat the permanent threat of flooding in the old districts of the capital; on the contrary, it automatically delegitimizes the traditional protection structures in the current city (stopping subsidence and firstly pumping) in favor of "heroic" engineering with significant potential political benefits for the Indonesian government [BAK 17, COL 17]. According to the project's opponents, it also has the disadvantage of blocking salt water from the agglomeration in the lagoon that is then transformed into an immense cesspool; NGOs condemn the utopia that consists of planning a closed lake of pure water, downstream from the sewers that the coastal rivers constitute today, even though water treatment stations are planned, which is enough to make the government hesitate. The Garuda project is therefore one of the global models of the return to colossal hydraulic infrastructure projects that have characterized the 20th Century, whereas softer means of approaching development are emerging. Remediation for the excesses of the hard engineering that was previously in fashion has reached the point of being forgotten, and the question has become how much capacity the semi-public consortium, which is partly directed by its private shareholders, has to transform its focus to create truly sustainable development and to be less subject to certain demonstrations of "scientific diplomacy".

At a more modest scale, the example of the infill of the Phnom Penh alluvial plain illustrates the will of many decision makers to remove, either partially or totally, the deltaic plains of the Mekong from the fluvial, or sea, areas. Another example is the

13 https://www.indonesia-investments.com/projects/public-private-partnerships/giant-sea-wall-jakarta-national-capital-integrated-coastal-development-ncicd/.

Camargue, where the Rhône Plan financed an increase in the height of the dykes in the 20th Century and aimed to protect all of the lands upstream from Vaccarès, to partially authorize urban development in the long term. The selected principle automatically denies the specificity of the deltaic plain, extends the main fluvial channel up to the mouth and confirms the status of the Camargue, which is a simple dyked alluvial plain. However, the Rhône floods in 1993, 1994, 2002 and 2003 demonstrate that the risk is always present and that a delta, an area of very shallow slope and high flow rates when in flood, can at any time regain its operational characteristics, just as a protected delta sinks, sometimes becomes a polder and is presented in a very vulnerable manner against the rise in sea levels.

5.5.2. *Some solutions for correction of the sedimentary deficit of deltaic plains*

Restoration of the deltas in the face of rising sea levels "implies diversion of sediments and of the water from the main channels towards the flooded areas to construct new lands and provide a platform for regeneration of ecosystems in wet environments" [PAO 11] in [GIO 13]. A few examples are of note.

5.5.2.1. *The lessons learned from diversion of the Mississippi*

We have seen that among the solutions that have been seen in the Mississippi Delta, diversion of waters loaded with sediments is considered to be credible, but it can only involve areas of a limited size, so less than 10% of the delta's surface (Chapter 2). Restoration of connections between the river and delta, reintroduction of water and fluvial sediments into the deltaic plain to restore marshes and aggradation of the ground, are recommended by many scientists and managers [DAY 16a]. Chapter 4 gave a reminder of the benefits in terms of ground height or restoration of organic soils that support vegetation which is then able to trap sediments. It is still necessary for the diversion process to be quite brutal, for the ground to quickly rise to compensate for the rise in sea levels. The costs of such operations are much lower than for dredging operations, because the energy at stake is natural and the beneficial effects can extend over a century, thus softening the investment. However, criticisms have taken root with regard to diversions: some of these include long-lasting flooding of low-lying areas, reduction of vegetation productivity in flooded marshes in the spring, increased rarity of commercial species in haline environments (shrimps, oysters, fish) to the benefit of less interesting freshwater species, but the improvement of productivity is, however, observed in the long term, thanks to plankton.

5.5.2.2. *The Rhône Delta*

The principle of resupply of low-lying areas in deltaic plains was proposed by the Mississippi experts in two deltas in the Western Mediterranean that were affected to a great degree, the Rhône Delta and the Ebro Delta. The Rhône Delta was cut off from its fluvial supply by dyking of the Grand Rhône and the Petit Rhône after the May 1856 flood, with the aim of turning over wetlands to agriculture in large typically latifundial areas; the Ebro Delta, supplied by small diversion channels between 1860 and 1966, was isolated by the construction of large reservoir-dams that blocked 99% of the solid load, whereas half of the flow rate is lost upstream (Chapter 2). Survival of these deltas occurs through their reconnection with the river or the sea to compensate for subsidence. Sites where entry of water, sediment minerals and nutrients are possible (input from the river, from the tide) do not lose ground to rising sea levels, because they conserve good productivity. However, if the salinity of marshes in connection with the sea increases too much, their agricultural productivity reduces, just like the organic production that becomes incapable of compensating for the rise in sea levels; we therefore find ourselves in a similar case to that of entirely disconnected areas. These principles plead in favor of maintaining natural processes [DAY 11]. Concretization of these principles was observed in the Camargue, where the 1993 and 1994 floods opened breaches in the left bank embankment of the Petit Rhône and created significant splay deposits over more than 100 km^2.

5.5.2.3. *The Ebro Delta*

The proposal has been made to remobilize the sediments that are trapped in the dams to take them towards the marshes and towards the rice paddies in order to aggrade the soils and increase fertilization. Before construction of the dams, was the net increase not on average 5 mm/year over 65% of the surface of the delta [PON 17]? The limit to implementation of flooding flow rates in the delta is the impossibility of taking sufficient discharge from the downstream dams, in the context of such high levels of control from upstream and due to the fact that the safety obtained by the hydraulic developments has encouraged rice production and urbanization and has therefore frozen the situation. Previously, we saw that flushing was only carried out on an experimental basis, but that it has instilled great hope for aggradation of soils through the use of irrigation and for at least counterbalancing the relative rise in sea levels. The results are a little disappointing, but the flushing is carried out when the fluvial channel is invaded by proliferations of algae and must be scoured[14].

14 José A. Jimenez, *in litteris*.

5.5.2.4. *The Danube Delta*

The short length of protection structures allows distributaries to supply a significant part of the deltaic plain with sediments. Measurements have revealed greater rates of deposition than expected, which correspond to more than 10% of the fluvial mineral flow (albeit reduced) that reaches the sea. However, the dams have reduced the sandy fraction of the flows entering the deltaic plain. In total, L. Giosan [GIO 13] is "cautiously optimistic" with regard to the future of a delta that has suffered from dams, but benefitted from incision of the channel since the 1950s. These channels have allowed more sediments to be diverted in the plain and towards the natural lakes than the natural branches did before the era of large construction works; they could be maintained to conserve their function supplying the delta's interior. On the other hand, the sedimentary deficit on the coast complicates the dynamics of this fringe that is reworked by waves and long-shore drift to the detriment of the old eroded lobes.

5.5.2.5. *The Brahmaputra Delta in Bangladesh*

A beneficial initiative is the Tidal River Management (TRM) project applied to the management of the risk of flooding in Bangladesh. Duly recording the siltation, the degradation of drainage of deltaic channels and saturation of polderized sectors with water in the 1960s, the populations themselves have proposed to reopen polders to restore the operation of estuaries in the district of Jessore, located between Calcutta and Padma, upstream from the estuaries that drain the Sundarbans. The TRM, practiced since the beginning of the decade 2000–2010, would be the best way to raise the ground, attenuate the water saturation crisis and protect the coastal region from the threat constituted by the rise in ocean levels [PAU 13].

5.5.3. *The sustainable solutions*

The inevitable and planned rise in global sea levels as well as the impossibility of protecting all deltaic terrains with structures built along the seafront means that structural defenses cannot be the only ones proposed. They cost too much for many deltas around the world and have proven undesirable effects. Soft, flexible and sustainable solutions are experienced around the world to increase the hazard resilience. UNISDR (United Nations Office for Disaster Risk Reduction) gives the following definition of resilience:

> "The ability of a system, community or society exposed to hazards to resist, absorb, accommodate, adapt to, transform and recover from the effects of a hazard in a timely and efficient manner, including through the preservation and restoration of its essential basic structures and functions through risk management" [UNI 09].

Given the loss of 700,000 human lives, as recorded between 2005 and 2015 in deltaic zones, for all types of natural disasters, improvement of the resilience is necessary. Just for the period 2008–2012, 144 million people were displaced, without taking into account the enormous financial losses (1,300 billion dollars).

5.5.3.1. *Risk education and vulnerability reduction*

Education is the primary factor in the protection of populations within deltas where no protection from cyclones and fluvial flooding is in place. The information, preparation for evacuation and a high level of reactivity are important. We have seen (Chapter 3) that, in Bangladesh, poverty can be so significant that the population remains on site, hoping to protect their property and livestock rather than saving their own lives. Bangladesh has improved its risk prevention with systems for warning, evacuation and protection (cyclone shelters), promotion of cooperation between communities and individuals, in such a way that the number of victims was greatly reduced to comparable situations in recent years.

On a global scale, after many international meetings and the initiatives of Platforms for Disaster Risk Reduction, the United Nations launched a 15-year program in 2015 [UNI 15]; this international strategy aims to look more in detail at the issue of vulnerability of zones subject to climatic risks. Other than understanding of the causes of risk and the improvement of governance, priority is given to improving resilience, to preparation, post-crisis recovery and reconstruction ("Rebuild better"); this applies to buildings in poor areas, the health system, the safety policy and many other subjects, as far as possible on a low financial budget.

5.5.3.2. *Developing knowledge and coordination*

Initiatives are surfacing which aim to found and coordinate initiatives on a global scale: thus, the idea of declaring the year 2013 the "International Year for Deltas" and using this to launch a Global Delta Sustainability Initiative [BRO 16a, FOU 13]. The idea is to have a worldwide partnership to save the world's deltas, again to "transform vulnerability into resilience" founded on pairing society and ecosystems in a unified system.

A promising initiative, which is not especially focused on delta management, is the International Water Stewardship Program (IWaSP); it attempts to combine the best practices identified in the field of water and local knowledge. The IWaSP works in Western and Southern Africa with public finances. It is managed by the Deutsche Gesellschaft für Internationale Zusammenarbeit (GIZ) in the name of the German Federal Ministry for Economic Cooperation and Development (BMZ) and the United Kingdom's Department for International Development (UKAID) that jointly finances it, with private companies like Coca-Cola. IWaSP contributes to sustainable water management through their expertise, in particular for use by the population.

5.5.3.3. *Adaptation of traditional production methods and organized retreat*

Conversion of agriculture is envisaged or in progress in the deltas that are experiencing an increase in salt intrusion. In the Mekong Delta, this occurs as a spontaneous conversion, from rice cultivation to shrimp farming.

Organized retreat means moving property and people that are under the greatest threat, with the consequence of abandoning part of the territory to natural dynamics, in other words, progressive retreat in the face of coastal erosion or internal submersion. This approach can occur through the conversion of terrestrial areas into marshes, as may be the case in the Camargue. Quite often, policies are passive, expressed by inaction in the face of increasing danger, either by choice or by obligation.

5.5.3.4. *Sea dams or natural defenses? The dilemma of the Netherlands*

Is it better to construct giant artificial barriers or to count on natural defenses to protect deltas? The debate began in 2014 on the basis of feedback from the Rhine-Meuse Delta and the Mississippi Delta [WES 14]. It is a novelty to be able to judge, in the Netherlands and once a situation of safety has been achieved, the environmental impacts of a large estuarian dam progressively constructed through successive trial and error.

The negative effects of correct management of natural resources and ecosystemic services:

– masses of freshwater that were initially held back behind the dam have been rapidly damaged by efflorescence of toxic algae due to large inputs of nutrients and made unsuitable for irrigation;

– the water bodies have caused stratification and anoxic* conditions, factors with a role in the mortality of fauna and flora;

– partial opening of the Eastern Scheldt barrier was intended to compensate for the impacts of a waterproof barrier on marine sediments. Since the barrier no longer allows sand to enter the estuaries, the strength of the tides has caused erosion of the mud flats and marshes and redistribution of sediments, all of this to the detriment of the stability of the dykes existing in the estuary and of the ecological value of the site (migratory birds were affected).

The position held by the Dutch Ministry for Transport moved in 2010 towards adaptative management, which gives precedence to low-cost and high-yield non-structural measures. These are structural measures, but they are known as "green": they must be capable of dissipating marine energy, guaranteeing continuity for sediments moving towards the sea, along with openings, exchanges and mixing of water bodies. In summary, robust, flexible and sustainable measures are required, which are capable of integrating environmental measures to manage changes that may not have been planned in the initial design of the very complex system. The managers seem to be focusing more on ambitious ecological engineering which is considered and tested in the same way as classic engineering.

5.5.3.5. *The Mekong Delta*

The CBDRM (Community-Based Disaster Risk Management in Vietnam) project was undertaken in 2007 by the Vietnamese government, in response to the serious damage suffered in 2006. The strategy, due to a lack of possible improvements of the situation created by the upper part of the basin, is to accompany the environmental changes by accepting them, as part of the program "Living Actively with Flooding". This project, for which the Netherlands were advisors, supported by many NGOs and co-financed by European countries, has several parts:

– taking advantage of the benefits generated by floods that bring nutrients, aquatic resources, freshwater fish, freshwater for human consumption, and which desalinate waters and wash soils containing acids or sulfates;

– reduction of the loss of human life due to flooding through adaptation of the residential habitat and risk management within communities. A prevention policy is put in place;

– reduction of flooding without containing it entirely. Floods will be retained in reservoirs by dams upstream, which can contain up to 10 km^3 and will intrude inland while being evacuated towards the Western sea and the Vam Co Ty. Further downstream, the pressure from the flood wave will be contained within large estuarian branches and in rice paddies through reduction of the floodplain;

– improvement of drainage is very difficult in the context of rising sea levels, so construction of large basins downstream of the delta is envisaged, of a total volume of 2 km^3. The basins, located at the mouths of the estuaries, will be contained within sea dykes that are designed to block intrusion of the sea during incoming tides and to retain fluvial flooding at high tide. Evacuation of the water will take place twice a day at low tide. This method of regulation will restrict saline intrusions and should guarantee a supply of drinking water, but doubts are justified about this subject;

– improvement of the management and condition of the remaining 65,000 ha of mangroves. Their role will be to be a buffer for the ocean, thanks to a width of 500–1,000 m in front of the sea dykes.

In a more general manner, sustainable management of tropical deltas therefore becomes a major issue for the scientific community and has a tendency to progressively take the upper hand against costly and disputable structural solutions from an environmental point of view [SZA 15].

Conclusion

The distinction made in the two volumes of this overview – on the one hand, the operation of the Earth's rivers in their watershed and, on the other hand, the deltas that they have constructed at their mouths in the seas and the oceans – is a convenience that may have excessively divided these two vast geographical objectives if we had not taken care to conserve a necessary level of coherence in the text. The three simplest possible conclusions, which now become necessary at the end of this overview, are that:

– the Earth's rivers no longer take on their former role in relation to deltas, a role which should have continued; they no longer supply them with sufficient amounts of sediment to compensate for their natural subsidence and even more so to allow them to conserve a configuration of equilibrium as they develop;

– at the same time, an unexpected phenomenon that was acknowledged later on – when the research community managed to take it into account – is that the deltas are threatened by the rapid rise in sea levels, in response to the composite effects of global warming. This rise is serious where mangroves have been weakened or eradicated;

– lastly, under threat from upstream continental factors and downstream oceanic factors, the deltas, or at least most of them, have been badly managed. Economic forces and often bad management originating from politics have combined to impose internal deterioration of the environmental conditions that are specific to deltas. Caught in the middle, deltas collapse inwards, under the extractions of hydrocarbons, water for mankind, irrigation, industry and the extractions carried out in the channels that aggravate subsidence and sedimentary deficit.

This three-part combination is of so great a magnitude, and the margins for maneuver are so much reduced in the three components of the crisis, that the only possible diagnosis is highly pessimistic. Bad management of rivers can be corrected, say the optimists. This is impossible to recognize when evaluating the construction of reservoir-dams on the last remaining free rivers and in the last watersheds, which are largely undeveloped. The pace of dam construction exceeds the pace of infilling of existing reservoirs, in such a way that the observed deficit in the annual budgets will continue at least for a long time. How can we stop, if reversal of the trend is impossible, the rise in sea levels? Optimists will say that applying the recommendations of the IPCC will be enough, but, in practice, concentrations of greenhouse gases are still increasing. It remains for us to ensure better management of the deltas, which seems simple to those who are optimistic, but we notice that, barring exceptions, the economy that preys on their resources is expanding and becoming worse.

How are deltas going to evolve in the decades to come? Those that pose the greatest problems are the large deltas with low potential (Mekong, Yangtze, Ganges–Brahmaputra, Mississippi, Magdalena, Ebro and Po) and unsustainable deltas (Tigris–Euphrates, Nile and Colorado) [DAY 16b]. Recommendations for management vary significantly from structural solutions to environmental solutions. We now know that structural solutions have the effect of providing effective protection for agricultural paddies and the rural population (or even the urban population), but subsiding paddies become polders under threat from high waters and deprived of natural fertilizers; these solutions are not sustainable, even though they fulfill positive functions in the short term, as is the case for the Ganges–Brahmaputra delta. With regard to environmental solutions, which advocate an ordered retreat of activities and of the society that is under threat from the sea, are they not a luxury of countries rich enough to relocate in favor of nature, more or less well reconstituted? Such is the case for the Mississippi, where soft solutions are gaining credibility, and also for the Rhine and Meuse Delta, where structural solutions must give way to (still timid) restoration of processes and environments. The beginning of a natural combat against the rise in ocean levels is displayed in the Mekong Delta near Ho Chi Minh City, a sign of the rapid progression of Vietnam's economy. However, a movement of this kind cannot be expressed in the urbanized deltas of Southeast Asia, where structural methods prevail, sometimes for the glory of cities of global importance.

Glossary

The glossary terms are marked with an asterisk * in the text.

Units

Gt/year: billion tons/year or 10^9 tons.

hm³: million m^3.

km³: billion m^3.

Ma: million years.

Mt: million tons.

Terms

Adsorption: fixation of ions in a solution on the colloids of soil.

Anoxic: depleted levels of dissolved oxygen in an aquatic environment.

Anthropocene: the epoch of the Holocene, defined by the significant impact of the actions of humanity. Its duration remains a source of controversy.

Avulsion: a brutal change in the course of a river that abandons its original course (see diffluence).

***Bayou*:** an old arm of the Mississippi in Louisiana and Arkansas (from a Native American word meaning "serpent" or "sinuosity").

Bed load: materials transported by tractive forces in the bottom of rivers.

BP: "before present", the present being conventionally regarded as the year 1950.

Cal BP: radiocarbon age calibrated by taking account of fluctuations over time of the levels of radiocarbon in the atmosphere.

Catchment (area): the surface drained by the branches of a hydrographic network.

Cheniers: sandy, mobile coastal bars formed by long-shore drift, sometimes abandoned in coastal marshes.

Coastline: the extreme limit of the waters on a coast.

Continental shelf: the subaqueous extension of the continent; it terminates some distance from the coast where the continental slope begins.

Continuum: succession of formations starting upstream and changing down a river; the term was created to describe ecological processes.

Crevasse splay: a fan-shaped deposit formed downstream of a breach created by a flood in a natural or artificial levee.

Delta (or Deltaic) lobe: river construction at the mouth of an active or former river channel.

Diffluence: change in the course of a waterway following a flood (see avulsion).

Distributary: a waterway originating from the primary canal or channel and taking part of its discharge.

Distribution channel: a delta channel that spreads out the liquid and the solid flow.

Ebb: current generated by the outgoing tide (ebb tide).

Embayment: a vast inlet in the form of a bay, filled with continental sediment and shaped by quaternary flows.

Estuary: flared mouth shaped by the tide. Certain deltas have distributors that fall into the category of estuaries.

Eustatism: fluctuations in sea level due to quaternary freeze-up and melting that retain or release continental water.

Fetch: the distance a wind has blown over the free surface of a body of water; it affects the height of waves.

Flocculation: agglomeration of fine particles in suspension to form flocs, which encourages deposition.

Flow: current generated by the rising tide.

Glacio-isostasy: see isostasy.

Glacio-isostatic adjustment: see isostasy.

***Grau*:** a natural opening to a barrier beach, allowing the flow of continental and marine waters (French term).

Holocene: the epoch of the quaternary spanning the last 10 millennia.

Hypoxic: characterized by a low oxygen content.

Intertidal zone: the coastal area between the high tide level and the low tide level.

Isostasy: movements to readjust the Earth's crust, due in this context to ice overloading on a continent and inversely to deglaciation.

***Lido*:** a coastal sandy shore closing in a lagoon. It results from long-shore drift.

Little Ice Age: the cold, wet period that spans from the 14th Century to the late 19th Century.

Long-shore drift: sediment displacement along a shoreline; the current is caused by wind, waves and the swell.

Macrophytes: aquatic plants visible to the naked eye.

Marine transgression: invasion of coastal areas of continents by the sea following a rise in sea levels or due to continental sinking.

Mega-dam: a dam with a height greater than or equal to 100 m and/or a storage capacity greater than or equal to 1 km^3. China had 115 mega-dams in 2008.

Microtidal: a sea with a small tidal range.

Mouth bar: a subaqueous, sandy-silty deposit, constructed immediately downstream of a delta channel mouth.

Oxbow lake: an oxbow-shaped lake, isolated from the river when a meander is cut off.

Permafrost: permanently frozen ground in the Earth's cold regions.

Progradation: advancement (of a delta).

***Ria*:** a waterway draining a low valley carved during a period of receding water levels and invaded by the sea during a period of rising water levels (Spanish term).

Ridges and swales: a series of sandy ridges and curved depressions that are formed in the inside of a developing meander.

Roughness: a characteristic of a channel which is able to slow the speed of the water. It depends on the shape of the bed and the nature of the particles making up the bed.

Sedimentary budget (or balance): assessment of sediment flow or flux into and out of a river system.

Sediment cells: sections of a coast with distinct behavior and sediment balance under the effect of input of materials from the continent and from long-shore drift.

Sediment production zone: the space subject to the processes of erosion feeding a river system.

Storm surge: a rise above normal sea level; for example, under the effect of an atmospheric depression, a storm or strong winds push the surface water towards the continent.

Subsidence: sinking of the Earth's crust.

Suspended materials: sediment, generally silt and clay, transported in a mass of water.

Thermoablation: permafrost is loosened by ice melt on the highest areas of relief and removed by surface flows.

Thermoabrasion: sea water, warmer than permafrost, loosens the foot of cliffs and facilitates wave action.

Thermokarst or cryokarst: the relief of cold regions characterized by the development of surface landforms associated with differential melting of the active layer of permafrost in summer (depressions and ground collapse events due to ground settling, often occupied by lakes). These processes occur as a result of permafrost ice melting in summer.

Tidal flat: coastal muddy marsh.

References

[ABD 13] ABDEL-SALAM D., "Le delta face au changement climatique", *Al Ahram*, no. 980, June 26, 2013.

[ABI 15] ABIDIN H.Z., ANDREAS H., GUMILAR I. *et al.*, "Study on the risk and impacts of land subsidence in Jakarta", *Proceedings of the IAHS*, no. 372, pp. 115–120, available at: https://www.proc-iahs.net/372/115/2015/, 2015.

[ADE 10] ADEGOKE J.O., OLOGUNORISA T.E., "An assessment of recent changes in the Niger Delta coastline using satellite imagery", *Journal of Sustainable Development*, vol. 3, no. 4, pp. 377–396, 2010.

[ADE 11] ADEKOLA O., MITCHELL G., "The Niger Delta wetlands: Threats to ecosystem services, their importance to dependent communities and possible management measures", *International Journal of Biodiversity Science, Ecosystem Services & Management*, vol. 7, no. 1, pp. 50–68, 2011.

[AHM 06] AHMAD K.Q., "Changement climatique, inondations et gestion des crues : Le cas du Bangladesh", *Hérodote*, no. 121, pp. 73–94, 2006.

[ALE 14] ALEXEEVA O., ROCHE Y., "Le point de vue chinois sur les enjeux des barrages chinois sur le Mékong", *L'Espace politique*, no. 24, available at: http://espacepolitique.revues.org/3268, 2014.

[ALL 98] ALLISON M.A., "Historical changes in the Ganges-Brahmaputra delta front", *Journal of Coastal Research*, vol. 14, no. 4, pp. 1269–1275, 1998.

[ANT 15] ANTHONY E.J., BRUNIER G., BESSET M. *et al.*, "Linking rapid erosion of the Mekong River delta to human activities", *Nature Scientific Reports*, no. 5, article no. 14745, 2015.

[ANT 17] ANTHONY E.J., BESSET M., DUSSOUILLEZ P., Recent shoreline changes and morpho-sedimentary dynamics of the Ayeyarwady River delta: Assessing the impact of anthropogenic activities on delta shoreline stability, Project, WWF, 2017.

[ARN 98] ARNAUD-FASSETTA G., Dynamiques fluviales holocènes dans le delta du Rhône, Thesis, Université Aix-Marseille I, 1998.

[ARN 02] ARNAUD-FASSETTA G., "Geomorphological records of a 'flood-dominated regime' in the Rhône Delta (France) between the 1st century BC and the 2nd century AD. What correlations with the catchment paleohydrology?", *Geodinamica Acta*, no. 15, pp. 79–92, 2002.

[ARN 03] ARNAUD-FASSETTA G., "River channel changes in the Rhône delta (France) since the end of the Little Ice Age: Geomorphological adjustment to hydroclimatic change and natural resource management", *Catena*, no. 51, pp. 141–172, 2003.

[ASI 00] ASIANICS AGRO DEVELOPMENT INTERNATIONAL LTD, Tarbela Dam and related aspects of the Indus River Basin Pakistan, Report for the World Commission on Dams, 2000.

[AUE 15] AUERBACH L.W., GOODBRED S.L. Jr., MONDAL D.R. *et al.*, "Flood risk of natural and embanked landscapes on the Ganges-Brahmaputra tidal delta plain", *Nature Climate Change*, no. 5, available at: http://dx.doi.org/10.1038/nclimate2472, 2015.

[AWO 13] AWORAWO D., "Deprivation and resistance: Environmental crisis, political action, and conflict resolution in the Niger Delta since the 1980s", *Journal of International and Global Studies*, vol. 4, no. 2, pp. 52–70, 2013.

[AYE 17] AYEB-KARLSSON S., "Facing disasters: Lessons from a Bangladeshi Island", *The Conversation*, available at: https://ehs.unu.edu/blog/ articles/16452.html, August 24, 2017.

[AZA 17] AZAD A., "Banishing refugees to a flood-prone island will not solve Bangladesh's Rohingya refugee crisis", *The Conversation*, available at: https://theconversation.com/banishing-refugees-to-a-flood-prone-island-will-not-solve-bangladeshs-rohingya-refugee-crisis-72853, March 27, 2017.

[BAK 17] BAKKER M., KISHIMOTO S., NOOY C., *Social Justice at Bay: The Dutch Role in Jakarta's Coastal Defence and Land Reclamation*, The Transnational Institute, Amsterdam, available at: https://www.tni.org/en/publication/social-justice-at-bay, 2017.

[BAL 91] BALESTRI M., VILLANI B., "Study of the subsidence in the Bolognese area", in JOHNSON A.I. (ed.), *Proceedings of the 4th International Symposium on Land Subsidence*, IAHS Publication no. 200, 1991.

[BAR 94] BARRAS J.A., BOURGEOIS P.E., HANDLEY L.R., Land loss in coastal Louisiana 1956–90, Report, National Biological Survey/National Wetlands Research Center Open File Report 94-01, 1994.

[BAR 97] BARRY J.M., *Rising Tide: The Great Mississippi Flood of 1927 and How it Changed America*, Simon & Schuster, New York, 1997.

[BAR 03] BARRAS J.A., BEVILLE S., BRITSCH D. *et al.*, Historical and projected coastal Louisiana land changes: 1978–2050, Report, Louisiana Coastal Area Comprehensive Coastwide Ecosystem Restoration Study, 2003 (revised in 2004).

[BAT 03] BATALLA R., "Sediment deficit in rivers caused by dams and instream gravel mining. A review with examples from NE Spain", *Revista Cuaternario & Geomorfologia*, vol. 17, nos 3–4, pp. 79–91, 2003.

[BAT 11] BATALLA R., VERICAT D., "An appraisal of the contemporary sediment yield in the Ebro Basin", *Journal of Soils Sediments*, no. 11, pp. 1070–1081, 2011.

[BEG 87] BEGUM K., *The Farakka Barrage: A Techno-Political Tangle in South-Asia*, The University Press Limited, Dhaka, 1987.

[BEN 14] BENEDIKTER S., "Extending the hydraulic paradigm: Reunification, state consolidation, and water control in the Vietnamese Mekong Delta after 1975", *Southeast Asian Studies*, vol. 3, no. 3, pp. 547–587, 2014.

[BER 07] BERG M., STENGEL C., TRANG P.T.K. *et al.*, "Magnitude of arsenic pollution in the Mekong and Red River Deltas – Cambodia and Vietnam", *Science of the Total Environment*, no. 372, pp. 413–425, 2007.

[BER 13] BERNIER J., Trends and causes of historical wetland loss in coastal Louisiana, Fact sheet 2013–3017, USGS, 2013.

[BES 17] BESSET M., ANTHONY E.J., DUSSOUILLEZ P. *et al.*, "The impact of Cyclone Nargis on the Ayeyarwady (Irrawaddy) River delta shoreline and nearshore zone (Myanmar): Towards degraded delta resilience?", *Comptes Rendus Geoscience*, no. 349, pp. 238–247, 2017.

[BHA 06] BHATTACHARYA J.P., "Deltas", *Facies Models Revisited, Society for Sedimentary Geology*, no. 84, pp. 237–292, 2006.

[BHA 13] BHATTACHARYA A.K., "An analysis of land subsidence in Bangkok and Kolkata due to over-extraction of groundwater", *Electronic Journal of Geotechnical Engineering*, no. 18, pp. 1683–1694, 2013.

[BID 17] BIDORN B., KISH S.A., DONOGHUE J.F. *et al.*, "Variability of the total sediment supply of the Chao Phraya River, Thailand", in WIEPRECHT S., HAUN S., WEBER K. *et al.* (eds), *River Sedimentation*, Taylor & Francis, London, 2017.

[BIG 09] BIGGS D., MILLER F., CHU T.H. *et al.*, "The delta machine: Water management in the vietnamese Mekong Delta in historical and contemporary perspectives", in MOLLE F., FORAN T., KAKÖKEN M. (eds), *Contested Waterscapes in the Mekong Region: Hydropower, Livelihoods and Governance*, Earthscan, London, 2009.

[BIN 05] BINH T.N.K.D., VROMANT N., HUNG N.T. *et al.*, "Land cover changes between 1968 and 2003 in Cai Nuoc, Ca Mau Peninsula, Vietnam", *Environment, Development and Sustainability*, no. 7, pp. 519–536, 2005.

[BLU 09] BLUM M.D., ROBERTS H.H., "Drowning of the Mississippi Delta due to insufficient sediment supply and global sea-level rise", *Nature Geoscience*, no. 27, pp. 488–491, 2009.

[BOG 17] BOGARD J., "In Bangladesh, people are eating more fish but getting less nutrition from it", *The Conversation*, April 11, 2017.

[BOL 15] BOLSHIYANOV D., MAKAROV A., SAVELIEVA L., "Lena River delta formation during the Holocene", *Biogeosciences*, no. 12, pp. 579–593, 2015.

[BRA 10] BRAMMER H., "After the Bangladesh flood action plan: Looking to the future", *Environmental Hazards*, no. 9, pp. 118–130, 2010.

[BRA 13a] BRAVARD J.-P., GOICHOT M., GAILLOT S., "Geography of sand and gravel mining in the Lower Mekong River. First survey and impact assessment", *Échogéo*, no. 26, pp. 2–18, available at: http://echogeo.revues.org/13659, December 19, 2013.

[BRA 13b] BRAVARD J.-P., GOICHOT M., TRONCHÈRE H., "An assessment of sediment-transport processes in the Lower Mekong River based on deposit grain sizes, the CM technique and flow-energy data", *Geomorphology*, no. 207, pp. 174–189, 2013.

[BRI 08] BRICHIERI-COLOMBI S., "Hydrological impact on Bangladesh of Chinese and Indian Plans on the Brahmapoutra", in MIRZA M.M.K., AHMEDA U., AHMAD Q.K. (eds), *Interlinking of Rivers in India: Issues and Concerns*, Taylor & Francis, Boca Raton, 2008.

[BRO 85] BROCHEUX P., *The Mekong Delta: Ecology, Economy and Revolution, 1860–1960*, Center for Southeast Asian Studies, Madison, 1985.

[BRO 15] BROWN S., NICHOLLS R.J., "Subsidence and human influences in megadeltas: The case of the Ganges-Brahmaputra-Meghna", *Science of the Total Environment*, nos 527–528, pp. 362–374, 2015.

[BRO 16a] BRONDIZIO E.S., FOUFOULA-GEORGIOU E., SZABO S. *et al.*, "Catalyzing action towards the sustainability of deltas", *Current Opinion in Environmental Sustainability*, no. 19, pp. 182–194, 2016.

[BRO 16b] BROWN J.M., PHELPS J.J.C., BARKWITH A. *et al.*, "The effectiveness of beach mega-nourishment, assessed over three management epochs", *Journal of Environmental Management*, no. 184, pp. 400–408, 2016.

[BUC 14] BUCKLEY M., *Meltdown in Tibet: China's Reckless Destruction of Ecosystems From the Highlands of Tibet to the Deltas of Asia*, Palgrave Macmillan, New York, 2014.

[BUD 15] BUDIYONE Y., AERTS J., BRINKMAN J. *et al.*, "Flood risk assessment for delta mega-cities: A case study of Jakarta", *Natural Hazards*, no. 75, pp. 389–413, 2015.

[CAM 07] CAMPANELLA R., Above sea level New Orleans: The residential capacity of Orleans parish's higher ground, White paper, Center for Bioenvironmental Research/Tulane and Xavier Universities, New Orleans, 2007.

[CAM 08] CAMPANELLA R., *Bienville's Dilemma: A Historical Geography of New Orleans*, Center for Lousiana Studies/University of Lousiana, Lafayette, 2008.

[CAP 91] CAPRA A., FOLLONI G., RUSSO P., "Subsidence controls in the town of Bologna", in JOHNSON A.I. (ed.), *Proceedings of the 4th International Symposium on Land Subsidence*, IAHS Publication no. 200, 1991.

[CAR 95] CARBOGNIN L., MARABINI F., TOSI L., "Land subsidence and degradation of the Venice littoral zone, Italy", in BARENDS F.B.J., BROUWER F.J.J., SCHRÖDER F.H. (eds), *Proceedings of the 5th International Symposium on Land Subsidence*, IAHS Publication no. 234, 1995.

[CAR 05] CARMINATI E., DOGLIONI C., SCROCCA D., "Magnitude and causes of long-term subsidence of the Po Plain and Venetian region", in FLETCHER C.A., SPENCER T., DA MOSTO J., CAMPOSTRINI P. (eds), *Flooding and Environmental Challenges for Venice and its Lagoon: State of Knowledge*, Cambridge University Press, Cambridge, 2005.

[CHA 06] CHAKRABORTY R.L., "Le delta du Bengale et la mise en valeur de terres alluviales depuis le régime colonial", *Hérodote*, no. 121, pp. 119–140, 2006.

[CHA 08] CHAO B.F., WU Y.H., LI Y.S., "Impact of artificial reservoir water impoundment on global sea level", *Science*, no. 320, pp. 212–214, 2008.

[CHA 11] CHANDIO N.H., ANWAR M.M., CHANDIO A.A., "Degradation of Indus delta, removal of mangroves forestland and its causes: A case study of Indus River delta", *Sindh University Research Journal (Science Series)*, vol. 43, no. 1, pp. 67–72, 2011.

[CHA 18] CHAPMAN A., TRI V.P.D., "Climate change is triggering a migrant crisis in Vietnam", *The Conversation*, available at: https:// theconversation.com/climate-change-is-triggering-a-migrant-crisis-in-vietnam-88791, January 19, 2018.

[CHE 12] CHEN Y., SYVITSKI J.P.M., GAO S. *et al.*, "Socio-economic impacts on flooding: A 4000-year history of the Yellow River, China", *Ambio*, no. 41, pp. 682–698, 2012.

[CHE 15] CHEN Y., OVEREEM I., KETTNER A.J. *et al.*, "Modeling flood dynamics along the superelevated channel belt of the Yellow River over the last 3000 years", *Journal of Geophysical Research Earth Surface*, no. 120, pp. 1321–1351, 2015.

[CHU 11] CHURCH J.A., WHITE N.J., "Sea-level rise from the late 19th to early 21st century", *Surveys in Geophysics*, vol. 32, nos 4–5, pp. 585–602, 2011.

[CHU 13] CHU J., "How The Netherlands became the biggest exporter of resilience?", *Fast Company*, available at: https://www.fastcompany.com/3020918/how-the-netherlands-became-the-biggest-exporter-of-resilience, January 11, 2013.

[COB 94] COBB J.C., *The Most Southern Place on Earth: The Mississippi Delta and the Roots of Regional Identity*, Oxford University Press, Oxford, 1994.

[COC 15] COCHRANE J., "What's clogging Jakarta's waterways? You name it", *The New York Times*, available at: https://www.nytimes.com/2016/10/04/world/asia/jakarta-indonesia-canals.html, October 3, 2015.

[COL 18] COLTEN C.E., "Raising New Orleans: Historical analogs and future environmental risks", *Environmental History*, no. 23, pp. 1–8, 2018.

[COL 17] COLVEN E., "Understanding the allure of big infrastructure: Jakarta's Great Garuda Sea Wall Project", *Water Alternatives*, vol. 10, no. 2, pp. 250–264, 2017.

[COS 06] COSTANZA R., DAY J.W., MITSCH W.J., "A new vision for New Orleans and the Mississippi delta: Applying ecological economics and ecological engineering", *Frontiers in Ecology and the Environment*, vol. 4, no. 9, pp. 465–472, 2006.

[CPR 17] COASTAL PROTECTION AND RESTORATION AUTHORITY (CPRA), Louisiana's comprehensive master plan for a sustainable coast committed to our coast, Report, State of Louisiana, available at: http://coastal.la.gov/our-plan/2017-coastal-master-plan/, 2017.

[CRO 12] CRONIN T.M., "Rapid sea-level rise", *Quaternary Science Reviews*, no. 56, pp. 11–30, 2012.

[DAI 16] DAI L., "Preventing and controlling land subsidence in Shanghai. Towards more integrated and effective land use and ground water governance in the Yangtze Delta", *Geophysical Research Abstracts*, vol. 18, article no. EGU2016-16451, 2016.

[DAL 14] DALBORN C., SCOTT A.H., LEWIS J.A., *Community Resettlement Prospects in Southeast Louisiana: A Multidisciplinary Exploration of Legal, Cultural, and Demographic Aspects of Moving Individuals and Communities*, The Tulane Institute on Water Resources Law & Policy, New Orleans, 2014.

[DAN 14a] DANDEKAR P., THAKKAR H., *Shrinking and Sinking Deltas: Major Role of Dams in Delta Subsidence and Effective sea Level Rise*, South Asia Network on Dams, Rivers and People, Delhi, 2014.

[DAN 14b] DANG V.K., DOUBRE C., WEBER C. *et al.*, "Recent land subsidence caused by the rapid urban development in the Hanoi region (Vietnam) using ALOS InSAR data", *Natural Hazards and Earth System Sciences*, no. 14, pp. 657–674, 2014.

[DAR 13] DARBY S.E., LEYLAND J., KUMMU M. *et al.*, "Decoding the drivers of bank erosion on the Mekong river: The roles of the Asian monsoon, tropical storms, and snowmelt", *Water Resources Research*, no. 49, pp. 2146–2163, 2013.

[DAR 16] DARBY S.E., HACKNEY C.R., LEYLAND J. *et al.*, "Fluvial sediment supply to a mega-delta reduced by shifting tropical-cyclone activity", *Nature*, no. 539, pp. 276–279, 2016.

[DAY 97] DAY J.W. Jr, MARTIN J.F., CARDOC L. *et al.*, "System functioning as a basis for sustainable management of deltaic ecosystem", *Coastal Management*, vol. 25, no. 2, pp. 115–153, available at: https://www.tandfonline.com/doi/abs/10.1080/089207597093 62315, 1997.

[DAY 07] DAY J.W. Jr., BOESCH D.F., CLAIRAIN E.J. *et al.*, "Restoration of the mississippi delta: Lessons from Hurricanes Katrina and Rita", *Science*, no. 315, pp. 1679–1684, 2007.

[DAY 11] DAY J.W., IBANEZ C., SCARTON F. *et al.*, "Sustainability of Mediterranean deltaic and lagoon wetlands with sea-level rise: The importance of river input", *Estuaries and Coasts*, no. 34, pp. 483–493, 2011.

[DAY 16a] DAY J.W., LANE R.R., D'ELIA C.F. *et al.*, "Large infrequently operated river diversions for Mississippi delta restoration", *Estuarine, Coastal and Shelf Science*, vol. 183, part B, pp. 292–303, available at: http://dx.doi.org/10.1016/ j.ecss.2016.05.001, 2016.

[DAY 16b] DAY J.W., AGBOOLA J., CHEN Z. *et al.*, "Approaches to defining deltaic sustainability in the 21st century", *Estuarine, Coastal and Shelf Science*, vol. 183, part B, pp. 275–291, 2016.

[DEF 57] DEFFONTAINES P., DURLIAT M., *Espagne du Levant*, Arthaud, Grenoble, 1957.

[DEL 11] DELAUNE R.D., WHITE J.R., "Will coastal wetlands continue to sequester carbon in response to an increase in global sea level? A case study of the rapidly subsiding Mississippi river deltaic plain", *Climatic Change*, no. 201, available at: http://www.evergladeshub.com/lit/pdf12/DeLaune12-ClimCh110.297-314-C-SequestWetlandsMississip.pdf, 2011.

[DUM 06] DUMAS P.G., *Les digues de la honte, ou les vautours du Bengale*, Publibook, Paris, 2006.

[ERB 14] ERBAN L.E., GORELICK S.M., ZEBKER H.A., "Groundwater extraction, land subsidence, and sea-level rise in the Mekong Delta, Vietnam", *Environmental Research Letters*, no. 9, 2014.

[ERB 16] ERBAN L.E., GORELICK S.M., "Closing the irrigation deficit in Cambodia: Implications for transboundary impacts on groundwater and Mekong River flow", *Journal of Hydrology*, no. 535, pp. 85–92, 2016.

[ERI 06] ERICSON J.P., VÖRÖSMARTY C.J., DINGMAN S.L. *et al.*, "Effective sea-level rise and deltas: Causes of change and human dimension implications", *Global and Planetary Change*, no. 50, pp. 63–82, 2006.

[FAB 14] FABRIS M., ACHILLI V., MENIN A., "Estimation of subsidence in Po Delta area (Northern Italy) by integration of GPS data, high-precision leveling and archival orthometric elevations", *International Journal of Geosciences*, no. 5, pp. 571–585, available at: http://www.scirp.org/journal/ijg, 2014.

[FAN 06a] FANCHETTE S., "Le delta du Nil : enjeux et limites du contrôle territorial par l'État", *Hérodote*, no. 2, pp. 165–189, 2006.

[FAN 06b] FANCHETTE S., "Le delta du Niger (Nigeria) : rivalités de pouvoir, revendications territoriales et exploitation pétrolière ou les ferments de la violence", *Hérodote*, no. 2, pp. 190–220, 2006.

[FAS 13a] FASMAN J., "Scratching a living", *The Economist*, available at: https://www.economist.com/united-states/2013/06/08/scratching-a-living, June 8, 2013.

[FAS 13b] FASMAN J., "Why are so many people leaving the Mississippi Delta?", *The Economist*, available at: https://www.economist. com/the-economist-explains/2013/10/20/why-are-so-many-people-leaving-the-mississippi-delta, October 20, 2013.

[FAS 16] FASULLO J.T., NEREM R.S., HAMLINGTON B., "Is the detection of accelerated sea level rise imminent?", *Nature Scientific Reports*, no. 6, article no. 31245, 2016.

[FIL 14] FILIP F., GIOSAN L., "Evolution of Chilia lobes of the Danube delta: Reorganization of deltaic processes under cultural pressures", *The Anthropocene*, no. 5, pp. 65–70, 2014.

[FIS 69] FISHER W.L., BROWN L.F., SCOTT A.J. *et al.*, *Delta Systems in the Exploration for Oil and Gas*, Bureau of Economic Geology, University of Texas at Austin, 1969.

[FIS 44] FISK H.N., *Geological Investigation of the Alluvial Valley of the Lower Mississippi River*, Mississippi River Commission, Vicksburg, 1944.

[FOU 13] FOUFOUA-GEORGIOU E., "A vision for a coordinated international effort on delta sustainability", in YOUNG G., PERILLO G.M.E. (eds), *Deltas: Landforms, Ecosystems and Human Activities, Proceedings of HP1, IAHS-IAPSO-IASPEI Assembly, Gothenburg, Sweden*, IAHS Publication no. 358, 2013.

[FRI 91] FRIHY O.E., KHAFAGY A.A., "Climate and induced changes in relation to shoreline migrations trends at the Nile delta promontories, Egypt", *Catena*, no. 18, pp. 197–211, 1991.

[FRI 03] FRIHY O.E., "The Nile Delta–Alexandria coast: Vulnerability to sea-level rise, consequences and adaptation", *Mitigation and Adaptation Strategies for Global Change*, no. 8, pp. 115–138, 2003.

[FRI 10] FRIHY O.E., DEABES E.A., SHEREET S.M. *et al.*, "Alexandria–Nile Delta coast, Egypt: update and future projection of relative sea-level rise", *Environmental Earth Science*, no. 61, pp. 253–273, 2010.

[FRI 15] FRINGS R.M., "Sand and gravel on the move: Human impacts on bed-material load along the Lower Rhine River", in HUDSON P., MIDDELKOOP H. (eds), *Geomorphic Approaches to Integrated Floodplain Management of Lowland Fluvial Systems in North America and Europe*, Springer, New York, 2015.

[FRO 02] FROMARD F., KIET L.C., "Les mangroves du Vietnam du Sud : Histoire récente, dynamique actuelle et perspectives", *Bois et Forêts des Tropiques*, no. 273, pp. 31–42, 2002.

[FUN 12] FUNABIKI A., SAITO Y., PHAI V.V. *et al.*, "Natural levees and human settlement in the Song Hong (Red River) delta, northern Vietnam", *The Holocene*, vol. 22, no. 6, pp. 637–648, 2012.

[FUR 09] FURUICHI T., WIN Z., WASSON R.J., "Discharge and suspended sediment transport in the Ayeyarwady River, Myanmar: Centennial and decadal changes", *Hydrological Processes*, no. 23, pp. 1631–1641, 2009.

[FUR 15] FURUNO K., KAGAWA A., KAZAOKA O. *et al.*, "Groundwater management based on monitoring of land subsidence and groundwater levels in the Kanto Groundwater Basin, Central Japan", *Proceedings of the IAHS*, no. 372, pp. 53–57, available at: https://www.proc-iahs.net/372/53/2015/, 2015.

[FUR 17] FURLONG K., KOOY M., "Worlding water supply: Thinking beyond the network in Jakarta", *International Journal of Urban and Regional Research*, vol. 41, no. 6, pp. 888–903, 2017.

[GAL 75] GALLOWAY W.E., "Process framework for describing the morphologic and stratigraphic evolution of deltaic depositional systems", in BROUSSARD M.L. (ed.), *Deltas, Models for Exploration*, pp. 87–98, Houston Geological Society, Houston, 1975.

[GAL 08] GALLOWAY G.E., "Flood risk management in the United States and the impact of Hurricane Katrina", *International Journal of River Basin Management*, vol. 6, no. 4, pp. 301–306, 2008.

[GAT 93] GATEWOOD W.B., WHAYNE J.M. (eds), *The Arkansas Delta: Land of Paradox*, The University of Arkansas Press, Fayetteville, 1993.

[GAU 17] GAUTHERET J., "À Venise, MOSE, le chantier maudit", *Le Monde*, available at: https://www.lemonde.fr/planete/article/2017/11/03/a-venise-mose-le-chantier-maudit_5209690_3244.html, November 4, 2017.

[GEN 11] GENDEREN R. (VAN), ROOD J., *Water Diplomacy: A Niche for the Netherlands?*, Clingendael Institute, The Netherlands, 2011.

[GEO 05] GEORGIOU I.Y., FITZGERALD D.M. *et al.*, "The impact of physical processes along the Louisiana Coast", *Journal of Coastal Research*, no. 44, pp. 72–89, 2005.

[GIL 06] GILARD O., "Risques d'inondation dans le delta du fleuve Rouge. De la nécessité d'améliorer leur prise en compte dans le processus d'aménagement du territoire", *Hérodote*, no. 121, pp. 42–54, 2006.

[GIO 05] GIOSAN L., DONNELLY J.P., VESPREMEANU E. *et al.*, "River delta morphodynamics: Examples from the Danube Delta", in GIOSAN L., BHATTACHARYA J.P. (eds), *River Deltas – Concepts, Models, and Examples*, SEPM, Special Publication, no. 83, pp. 393–411, 2005.

[GIO 13] GIOSAN L., CONTANTINESCU S., FILIP F. *et al.*, "Maintenance of large deltas through channelization: Nature vs. humans in the Danube delta", *Anthropocene*, no. 1, pp. 35–45, 2013.

[GIO 14] GIOSAN L., SYVITSKI J., DAY J., "Protect the world's deltas", *Nature*, no. 516, pp. 31–33, 2014.

[GIO 17] GIOSAN L., NAING T., TUN M.M. *et al.*, "On the Holocene evolution of the Ayeyawady Megadelta", *Earth Surface Dynamics*, no. 6, pp. 451–466, available at: https://doi.org/10.5194/esurf-2017-64, 2017.

[GOR 85] GORDON R., "The Irawadi River", *Proceedings of the Royal Geographical Society and Monthly Record of Geography*, no. 7, pp. 292–331, 1885.

[GRA 05] GRAMLING R., HAGELMAN R., "A working coast: People in the Louisiana Wetlands", *Journal of Coastal Research*, no. 44, pp. 112–133, 2005.

[GRU 13] GRUMBINE R.E., PANDIT M.K., "Threats from India's Himalaya Dams", *Science*, no. 339, pp. 36–37, 2013.

[GUN 11] GUNAWAN T., "Challenges in creating sustainable urban polder in developing countries, case study: Development of Pluit polder, Jakarta", *The 5th Conference of the International Forum on Urbanism*, National University of Singapore, Singapore, 2011.

[GUP 12] GUPTA H., KAO S., DAI M., "The role of mega dams in reducing sediment fluxes: A case study of large Asian rivers", *Journal of Hydrology*, nos 464–465, pp. 447–458, 2012.

[HAN 10] HANSEN J., RUEDY R., SATO M. *et al.*, "Global surface temperature change", *Reviews of Geophysics*, no. 48, pp. 1–29, 2010.

[HAN 13a] HANAFY R., "Ressources naturelles : Le delta menacé d'immersion", *Al Arham*, January 30, 2013.

[HAN 13b] HANEBUTH T.J.J., KUDRASS H.R., LINSTÄDTER J. *et al.*, "Rapid coastal subsidence in the central Ganges-Brahmaputra Delta (Bangladesh) since the 17th century deduced from submerged salt-producing kilns", *Geology*, vol. 41, no. 9, pp. 987–990, 2013.

[HED 10] HEDLEY P.J., BIRD M.I., ROBINSON R.A.J., "Evolution of the Irrawaddy delta region since 1850", *Geographical Journal*, no. 176, pp. 138–149, 2010.

[HES 02] HESSELINK A.W. (ed.), "History Makes a River. Morphological changes and human interference in the river rhine, the Netherlands", *Nederlandse, Geografische Studies*, no. 292, 2002.

[HIG 13] HIGGINS S., OVEREEM I., TANAKA A. *et al.*, "Land subsidence at aquaculture facilities in the Yellow River Delta, China", *Geophysical Research Letters*, no. 40, pp. 3898–3902, 2013.

[HOA 07] HOA L.T.V., NHAN N.H., WOLANSKI E. *et al.*, "The combined impact on the flooding in Vietnam's Mekong River delta of local man-made structures, sea level rise, and dams upstream in the river catchment", *Estuarine, Coastal and Shelf Science*, no. 71, pp. 110–116, 2007.

[HOC 12] HOC D.X., "Mekong river delta: Challenges and solutions in terms of climate change and sea level rise", *4th International Conference on Estuaries and Coasts*, Water Resources University, Hanoi, Vietnam, October 8–11, 2012.

[HOL 08] HOLLAND R.A., *The Mississippi River in Maps and Views, From Lake Itasca to the Gulf of Mexico*, Rizzoli International Publications, New York, 2008.

[HOL 11] HOLM G.O. Jr., EVERS E., SASSER C.E., *The Nutria in Louisiana: A Current and Historical Perspective*, The Lake Pontchartrain Basin Foundation, New Orleans, 2011.

[HOR 07a] HORI K., SAITO Y., "Classification, architecture, and evolution of large-river deltas", in GUPTA A. (ed.), *Large Rivers: Geomorphology and Management*, Wiley & Sons, Chichester, 2007.

[HOR 07b] HORI K., SAITO Y., "An early Holocene sea-level jump and delta initiation", *Geophysical Research Letters*, vol. 34, no. 18, article no. L18401, 2007.

[HOR 10] HOROWITZ A.J., "A quarter century of declining suspended sediment fluxes in the Mississippi River and the effect of the 1993 flood", *Hydrological Processes*, vol. 24, no. 1, pp. 13–34, 2010.

[HUN 11] HUNG N.N., DELGADO J.M., TRI V.K. *et al.*, "Floodplain hydrology of the Mekong Delta, Vietnam", *Hydrological Processes*, no. 26, 2011.

[IBA 97] IBANEZ C., CANICIO A., DAY J. *et al.*, "Morphologic development, relative sea level rise and sustainable management of water and sediment in the Èbre Delta, Spain", *Journal of Coastal Conservation*, no. 3, pp. 191–202, 1997.

[IBE 96] IBE A.C., "The Niger Delta and sea-level rise", in MILLIMAN J.D., HAQ B.U. (eds), *Sea-Level Rise and Coastal Subsidence*, Kluwer Academic Publishers, pp. 249–267, 1996.

[IKE 09] IKEJIAKU B.V., "The concept "Poverty" towards understanding in the context of developing countries "*Poverty qua Poverty*": With some comparative evidence on Britain", *Journal of Sustainable Development*, vol. 2, no. 2, pp. 3–13, 2009.

[ISL 99] ISLAM M.R., BEGUM S.F., YAMAGUSHI Y. *et al.*, "The Ganges and the Brahmapoutra rivers in Bangladesh: Basin denudation and sedimentation", *Hydrological Processes*, vol. 13, no. 17, pp. 2907–2923, 1999.

[ITE 13] ITE A.E., IBOK U.J., ITE M.U. *et al.*, "Petroleum exploration and production: Past and present environmental issues in the Nigeria's Niger Delta", *American Journal of Environmental Protection*, vol. 1, no. 4, pp. 78–90, 2013.

[JIM 11] JIMENEZ J.A., VALDEMORO H.I., BOSOM E. *et al.*, "Storm-induced coastal hazards in the Ebro delta (NW Mediterranean)", *The Proceedings of the Coastal Sediments 2011*, pp. 1332–1345, 2011.

[JOH 13] JOHNSON T., "Sinking cities: Why rising seas threaten many of the world's great destinations", *The Weather Channel*, September 12, 2013.

[KAD 12] KADAFA A.A., "Environmental impacts of oil exploration and exploitation in the Niger Delta of Nigeria", *Global Journal of Science Frontier Research, Environment & Earth Sciences*, vol. 12, no. 3, pp. 18–28, 2012.

[KAK 08] KÄKÖNEN M., "Mekong Delta at the crossroads: More control or adaptation?", *Ambio*, vol. 37, no. 3, pp. 205–212, 2008.

[KEE 16] KEESSEN A., VINK M.J., WIERING M. *et al.*, "Solidarity in water management", *Ecology and Society*, vol. 21, no. 4, available at: https://doi.org/ 10.5751/ES-08874-210435, 2016.

[KEL 03] KELMAN A., *A River and its City: The Nature of Landscape in New Orleans*, University of California Press, Oakland, 2003.

[KES 03] KESEL R.H., "Human modifications to the sediment regime of the Lower Mississippi River flood plain", *Geomorphology*, no. 56, pp. 325–334, 2003.

[KHA 15] KHALIL S.M., FREEMAN A.M., "Challenges of ecosystem restoration in Louisiana. Availability of sediment and its management", in XU Y.J., ALLISON M.A., BENTLEY S.J. *et al.* (eds), *Sediment Dynamics from the Summit to the Sea, Proceedings of a symposium held in New Orleans, Louisiana, USA, 11–14 December 2014*, IAHS Publication no. 367, 2015.

[KIM 17] KIMMELMAN M., "Jakarta is sinking so fast, it could end up underwater", *The New York Times*, December 21, 2017.

[KOE 12] KOEHNKEN L., Potential sediment contribution in the Lower Mekong River Basin based on GIS analyses and sediment monitoring results, Contribution to the IWRM technical report on significant tributaries to the Mekong River System, IKPM report, MRC, 2012.

[KOM 12] KOMORI D., NAKAMURA S., KIGUCHI M. *et al.*, "Characteristics of the 2011: Chao Phraya River flood in Central Thailand", *Hydrological Research Letters*, no. 6, pp. 41–46, 2012.

[KON 14] KONDOLF G.M., RUBIN Z.K., MINEAR J.T., "Dams on the Mekong: Cumulative sediment starvation", *Water Resources Research*, no. 50, 2014.

[KOT 00] KOTB T.H.S., WATANABE T., OGINO Y. *et al.*, "Soil salinization in the Nile Delta and related policy issues in Egypt", *Agricultural Water Management*, no. 43, pp. 239–261, 2000.

[KUE 13] KUENZER C., VO QUOC T., "Assessing the ecosystem services value of Can Gio mangrove biosphere reserve: Combining earth observation and household-survey-based analyses", *Applied Geography*, no. 45, pp. 167–184, 2013.

[KUM 10] KUMMU M., LU X.X., WANG J.J. *et al.*, "Basin-wide sediment trapping efficiency of emerging reservoirs along the Mekong", *Geomorphology*, no. 119, pp. 181–197, 2010.

[LAN 04] LANDURE C., PASQUALINI M. (eds), *Supplément Bulletin Archéologique de Provence Volume 2: Delta du Rhône*, Association Provence archéologie, 2004.

[LAS 05] LASKA S., WOODDELL G., HAGELMAN R. *et al.*, "At risk: The human, community and infrastructure resources of Coastal Louisiana", *Journal of Coastal Research*, no. 44, pp. 90–111, 2005.

[LEW 76] LEWIS P., *New Orleans: The Making of an Urban Landscape*, Ballinger Publishing, Cambridge, 1976.

[LEW 18] LEWIS P., *New Orleans: The Making of an Urban Landscape,* 2nd edition, University of Virginia Press, Charlottesville, 2018.

[LI 17] LI X., LIU J.P., SAITO Y. *et al.*, "Recent evolution of the Mekong Delta and the impacts of dams", *Earth-Science Reviews*, no. 175, pp. 1–17, 2017.

[LIE 16] LIEW S.C., GUPTA A., CHIA A.S. *et al.*, "The flood of 2011 in the lower Chao Phraya valley, Thailand: Study of a long-duration flood through satellite images", *Geomorphology*, no. 262, pp. 112–122, 2016.

[LOI 14] LOISEL H., MANGIN A., VANTREPOTTE V. *et al.*, "Variability of suspended particulate matter concentration in coastal waters under the Mekong's influence from ocean color (MERIS) remote sensing over the last decade", *Remote Sensing of Environment*, no. 150, pp. 218–230, 2014.

[LUO 17] LUO X.X., YANG S.L., WANG R.S. *et al.*, "New evidence of Yangtze delta recession after closing of the Three Gorges Dam", *Nature Scientific Reports*, no. 7, article no. 41735, 2017.

[MAI 07] MAILLET G.M., RACCASI G., PROVANSAL M. *et al.*, "Transferts sédimentaires dans le Bas-Rhône depuis le milieu du XIXe siècle : essai de quantification", *Géographie physique et Quaternaire*, vol. 61, no. 1, pp. 39–53, 2007.

[MAL 12] MALEK A., "Some aspects of alluvial land reclamation process in Bangladesh during the colonial rule", *Vidyasagar University Journal of History*, no. 1, pp. 205–22, 2012–2013.

[MAN 15] MANH N.V., DUNG N.V., HUNG N.N. *et al.*, "Future sediment dynamics in the Mekong Delta floodplains: Impacts of hydropower development, climate change and sea level rise", *Global and Planetary Change*, no. 127, pp. 22–33, 2015.

[MAR 88] MARSH P., Lakes and Water in the Mackenzie Delta, National Hydrology Research Institute, Aurora Research Institute Report no. 6, 1988.

[MAR 89] MARSH P., HEY M., "The flooding hydrology of Mackenzie Delta lakes near Inuvik, N.W.T., Canada", *Arctic*, vol. 42, no. 1, pp. 41–49, 1989.

[MAR 08] MARET I., GOEURY R., "La Nouvelle-Orléans et l'eau : un urbanisme à haut risque", *Environnement urbain*, no. 2, available at: http://eue.revues. org/867, 2008.

[MAR 12] MARRINER N., FLAUX C., KANIEWSKI D. *et al.*, "ITCZ and ENSO-like pacing of Nile delta hydro-geomorphology during the Holocene", *Quaternary Science Reviews*, no. 45, pp. 73–84, 2012.

[MAR 13] MARRINER N., FLAUX C., MORHANGE C. *et al.*, "Tracking Nile Delta vulnerability to Holocene change", *PLOS ONE*, vol. 8, no. 7, article no. e69195, 2013.

[MEU 07] MEULEN M. (VAN DER), SPECK J.F. (VAN DER), LANGE G. (DE) *et al.*, "Regional sediment deficits in the Dutch Lowlands: Implications for long-term land-use options", *Journal of Soils and Sediments*, vol. 7, no. 1, pp. 9–16, 2007.

[MEZ 16] MEZZI P., "The Great Garuda, the masterplan to save Jakarta", *Abitare*, available at: http://www.abitare.it/en/habitat-en/urban-design-en/2016/01/24/the-great-garuda-the-masterplan-to-save-jakarta/, January 24, 2016.

[MID 01] MIDDLEKOOP H., DAAMEN K., GELLENS D. *et al.*, "Impact of climate change on hydrological regimes and water resources management in the Rhine basin", *Climatic Change*, no. 49, pp. 105–128, 2001.

[MIL 13] MILLIMAN J.D., FARNSWORTH K.L., *River Discharge to the Coastal Ocean. A Global Synthesis*, Cambridge University Press, Cambridge, 2013.

[MIL 87] MILLIMAN J.D., QIN Y.S., REN M.-E. *et al.*, "Man's influence on the erosion and transport of sediment by Asian rivers: The Yellow River (Huanghe) example", *Journal of Geology*, no. 95, pp. 751–762, 1987.

[MIT 01] MITSCH W.J., DAY J.W. Jr., GILLIAM J.W. *et al.*, "Nitrogen loading to the Gulf of Mexico from the Mississippi River Basin: Strategies to counter a persistent ecological problem", *Bioscience*, vol. 51, no. 5, pp. 373–388, 2001.

[MOL 06] MOLLE F., "Géographie de l'eau et interactions scalaires : le delta du Chao Phraya (Thaïlande) et son bassin versant", *Hérodote*, no. 121, pp. 141–164, 2006.

[MOR 02] MORRIS J.T., SUNDARESHWAR P.V., NIETCH C.T. *et al.*, "Responses of coastal wetlands to rising sea level", *Ecology*, vol. 83, no. 10, pp. 2869–2877, 2002.

[MOS 16] MOSSA J., "The changing geomorphology of the Atchafalaya River, Louisiana: A historical perspective", *Geomorphology*, no. 252, pp. 112–127, 2016.

[MRC 07] MISSISSIPPI RIVER COMMISSION (MRC), The Mississippi River & Tributaries Project: Designing the Project Flood, Information Paper, MRC, 2007.

[MUL 11] MULDER J.P.M., HOMMES S., HORSTMAN E.M., "Implementation of coastal erosion management in the Netherlands", *Ocean and Coastal Management*, no. 54, pp. 888–897, 2011.

[MUS 14] MUSA Z.N., POPESCU I., MYNETT A., "The Niger Delta's vulnerability to river floods due to sea level rise", *Natural Hazards and Earth System Sciences*, no. 14, pp. 3317–3329, 2014.

[NCI 14] NCICD CONSORTIUM, NCICD Master Plan, National Capital Integrated Coastal Development, Jakarta, 2014.

[NDC 93] NETHERLANDS DEVELOPMENT COOPERATION (NDC), Flood Action Plan, Bangladesh, Assessment Report, Ministry of Foreign Affairs, 1993.

[NGU 95] NGUYEN T.Q., HELM D.C., "Land subsidence due to groundwater withdrawal in Hanoi, Vietnam", in BARENDS F.B.J., BROUWER F.J.J., SCHRÖDER F.H. (eds), *Land Subsidence, Proceedings of the 5th International Symposium on Land Subsidence*, IAHS Publication no. 234, 1995.

[NGU 16] NGUYEN N.T., "Mekong delta: Past and current challenges awareness and actions", *Conference on Sustainable Uses of the Mekong Water Resources*, Can Tho, Vietnam, April 22–23, 2016.

[NIC 92] NICOLINO F., "Bangladesh, les digues de la honte", *Politis*, no. 6, 1992.

[NIE 01] NIENHUIS P.H., LEUVEN S.E.W., "River restoration and flood protection: Controversy or synergism?", *Hydrobiologia*, no. 444, pp. 85–99, 2001.

[NWI 06] NWILO P.C., BADEJO O.T., "Impacts and management of oil spill pollution along the Nigerian coastal areas", *Administring Marine Spaces: International Issues*, no. 119, 2006.

[ODO 11] ODOEMENE A., "Social consequences of environmental change in the Niger Delta of Nigeria", *Journal of Sustainable Development*, vol. 4, no. 2, pp. 123–135, 2011.

[OLU 12] OLUKAYODE A.J., "The Niger Delta Environmental Crisis in Nigeria: A perspective analysis", *African Research Review*, vol. 6, no. 3, pp. 150–164, 2012.

[OVI 10] OVIASUYI P.O., UWADIE J., "The dilemma of Niger-Delta Region as oil producing states of Nigeria", *Journal of Peace, Conflict and Development*, no. 16, pp. 110–126, 2010.

[OYE 16] OYEGUN C.U., LAWAL O., OGORO M., "The dynamic nature of Niger Delta shoreline", *IOSR Journal of Environmental Science, Toxicology and Food Technology*, vol. 10, no. 5, pp. 50–58, 2016.

[PAN 98] PANIN N., JIPA D., "Danube River sediment input and its interaction with the north-western Black Sea: Results of EROS-200 and EROS-21 projects", *Geo-Eco-Marina*, no. 3, pp. 23–35, 1998.

[PAN 12] PANIN N., OVERMARS W., "The Danube delta evolution during the Holocene: Reconstruction attempt using geomorphological and geological data, and some of the existing cartographic documents", *Geo-Eco-Marina*, no. 18, pp. 75–104, 2012.

[PAO 11] PAOLA C., TWILLEY R.R., EDMONDS D.A. *et al.*, "Natural processes in delta restoration: Application to the Mississippi Delta", *Annual Review of Marine Science*, no. 3, pp. 67–91, 2011.

[PAU 13] PAUL A., NATH B., ABBAS R., "Tidal River Management (TRM) and its implication in disaster management: A geospatial study on Hari-Teka river basin, Jessore, Bangladesh", *International Journal of Geomatics and Geosciences*, vol. 4, no. 1, 2013.

[PET 14] PETEUIL C., FRETAUD T., WIRZ C. *et al.*, "Importance of field observations for managing sediment fluxes in hydropower projects design and operation", *Proceedings of the 19th IAHR-APD Congress*, Hanoi, Vietnam, 2014.

[PHI 06] PHIEN-WEJ N., GIAO P.H., NUTALAYA P., "Land subsidence in Bangkok, Thailand", *Engineering Geology*, no. 82, pp. 187–201, 2006.

[PHI 17] PHILIP B., "Djakarta, ville d'extrêmes", *Le Monde*, 30 August 2017.

[PHO 95] PHO N.V., TUAN V.V., *The Mekong Delta Natural Conditions*, Agriculture Press, Hanoi, 1995.

[PIC 88] PICON B., *L'espace et le temps en Camargue*, Actes Sud, Arles, 1988.

[PON 17] PONT D., DAY J.W., IBANEZ C., "The impact of two large floods (1993–1994) on sediment deposition in the Rhône delta: Implications for sustainable management", *Science of the Total Environment*, no. 609, pp. 251–262, 2017.

[POW 13] POWELL L.N., *The Accidental City: Improvising New Orleans*, Harvard University Press, Cambridge, 2013.

[PRO 02] PROVANSAL M., BRUNETON H., VELLA C. *et al.*, "Paléohydrologie holocène dans la basse vallée du Rhône, d'Orange à la mer", in BRAVARD J.-P., MAGNY M. (eds), *Les paléoenvironnements fluviatiles et lacustres en France depuis 15 000 ans*, Errance, Paris, 2002.

[PRO 14] PROVANSAL M., DUFOUR S., SABATIER F. *et al.*, "The geomorphic evolution and sediment balance of the lower Rhône River (Southern France) over the last 130 years: Hydropower dams versus other control factors", *Geomorphology*, no. 219, pp. 27–41, 2014.

[RAM 15] RAMACHANDRAN S., "Water wars: China, India and the great dam rush", *The Diplomat*, available at: https://thediplomat.com/2015/ 04/water-wars-china-india-and-the-great-dam-rush/, April 3, 2015.

[RAM 18] RAMMELT C.F., MASUD Z.M., MASUD A., "The waterways of Tangail: Failures to learn from flood-control efforts in the Brahmaputra Basin of Bangladesh", *Water Alternatives*, vol. 11, no. 1, pp. 106–124, 2018.

[RAS 12] RASUL G, MAHMOOD A., SADIQ A. *et al.*, "Vulnerability of the Indus Delta to climate change in Pakistan", *Pakistan Journal of Meteorology*, vol. 8, no. 16, pp. 89–107, 2012.

[REN 13] RENAUD F.G., SYVITSKI J.P.M., SEBESVARIZ Z. *et al.*, "Tipping from the Holocene to the Anthropocene: How threatened are major world deltas?", *Current Opinion in Environmental Sustainability*, no. 5, pp. 1–11, 2013.

[REU 04] REUSS M., *Designing the Bayous: The Control of Water in The Atchafalaya Basin (1800–1995)*, Texas A & M University Press, College Station, 2004.

[RIG 18] RIGAUD K.K., SHERBININ A. (DE), JONES B. *et al.*, Groundswell: Preparing for Internal Climate Migration, Report, The World Bank, Washington D.C., 2018.

[RMS 07] RISKS MANAGEMENT SOLUTIONS (RMS), The 1927 Great Mississippi Flood: 80-years retrospective, Special Report, RMS, 2007.

[ROB 07] ROBINSON R.A.J., BIRD M.I., OO N.W. *et al.*, "The Irrawaddy river sediment flux to the Indian ocean: The original nineteenth-century data revisited", *Journal of Geology*, no. 115, pp. 629–640, 2007.

[ROV 07] ROVIRA A., IBANEZ C., "Sediment management options for the Lower Ebro River and its delta", *Journal of Soils and Sediments*, vol. 7, no. 5, pp. 285–295, available at: http://dx.doi.org/10.1065/jss2007.08.244, 2007.

[ROY 10] THE ROYAL SOCIETY & THE AMERICAN ASSOCIATION FOR THE ADVANCEMENT OF SCIENCE, *New Frontiers in Science Diplomacy*, The Royal Society, London, 2010.

[RUF 18] RUFFINI P.-B., "La diplomatie climatique ou comment la science peut servir les intérêts nationaux", *The Conversation*, February 11, 2018.

[SAI 01] SAITO Y., YANG Z., HORI K., "The Huanghe (Yellow River) and Changjiang (Yangtze River) deltas: A review on their characteristics, evolution and sediment discharge during the Holocene", *Geomorphology*, no. 41, pp. 219–231, 2001.

[SAI 05] SAIKKU M., *This Delta, This Land: An Environmental History of the Yazoo-Mississippi Floodplain*, The University of Georgia Press, Athens, 2005.

[SAI 07] SAITO Y., CHAIMANEE N., JARUPONGSAKUL T. *et al.*, "Shrinking megadeltas in Asia: Sea-level rise and sediment reduction impacts from case study of the Chao Phraya Delta", *LOICZ Inprint*, no. 2, pp. 3–9, 2007.

[SAT 06] SATO C., HAGA M., NISHINO J., "Land subsidence and groundwater management in Tokyo", *International Review for Environmental Strategies*, vol. 6, no. 2, pp. 403–424, 2006.

[SAX 27] SAXON L., *Father Mississippi*, The Century Company, New York, 1927.

[SCH 15] SCHMIDT C.W., "Delta subsidence: An imminent threat to coastal populations", *Environmental Health Perspectives*, vol. 123, no. 8, pp. 204–209, 2015.

[SHA 09] SHAJAHAN M., "End-user perspectives of multi-jurisdictional river basins: South Australia and Bangladesh", *South Australian Geographical Journal*, no. 108, pp. 38–58, 2009.

[SHU 93] SHU L., FINLAYSON B., "Flood management on the lower Yellow River: Hydrological and geomorphological perspectives", *Sedimentary Geology*, no. 85, pp. 285–296, 1993.

[SIO 28] SION J., *Géographie universelle Volume 4 : Asie des Moussons*, Librairie Armand Colin, Paris, 1928.

[SMI 11] SMITH D.E., HARRISON S., FIRTH C.R., "The early Holocene sea level rise", *Quaternary Science Reviews*, no. 30, pp. 1846–1860, 2011.

[SOL 07] SOLOMON S., QIN D., MANNING M. *et al.* (eds), *Climate change 2007, the physical science basis, Contribution of Working Group I to the Fourth Assessment Report of the Intergovernmental Panel on Climate Change*, IPCC, Cambridge University Press, Cambridge, 2007.

[SOO 11] SOOD A., MATHUKUMALLI B.K.P., "Managing international river basins: Reviewing India–Bangladesh transboundary water issues", *International Journal of River Basin Management*, vol. 9, no. 1, pp. 43–52, 2011.

[SPA 14] SPANGER-SIEGFRIED E., FITZPATRICK M.F., DAHL D., Encroaching Tides: How Sea Level Rise and Tidal Flooding Threaten U.S. East and Gulf Coast Communities Over the Next 30 Years, Report, Union of Concerned Scientists, Cambridge, 2014.

[STA 40] STAMP L.D., "The Irrawaddy River", *The Geographical Journal*, no. 95, pp. 329–352, 1940.

[STA 93a] STANLEY D.J., WARNE A.G., "Sea level and intitiation of Predynastic culture in the Nile delta", *Nature*, no. 363, pp. 435–38, 1993.

[STA 93b] STANLEY D.J., WARNE A.G., "Nile Delta: Recent geological evolution and human impacts", *Science*, no. 260, pp. 628–634, 1993.

[STA 94] STANLEY D.J., WARNE A.G., "Worldwide initiation of Holocene deltas by deceleration of sea-level rise", *Science*, no. 265, pp. 228–231, 1994.

[STE 13] STECKLER M.S., MONDAL D.R., NOONER S.L. *et al.*, GPS velocity field in Bangladesh: Delta subsidence, seasonal water loading and shortening across the Burma Accretionary Prism and Shillong Massif, *Fall Meeting 2013*, AGU American Geophysical Union, San Francisco, United States, 2013.

[SYV 05] SYVITSKI J.P.M., VÖRÖSMARTY C.J., KETTNER A.J. *et al.*, "Impact of humans on the flux of terrestrial sediment to the global coastal ocean", *Science*, no. 308, pp. 376–380, 2005.

[SYV 09] SYVITSKI J.P.M., KETTNER A.J., OVEREEM I. *et al.*, "Sinking deltas due to human activities", *Nature Geoscience*, no. 2, pp. 681–686, 2009.

[SYV 11] SYVITSKI J.P.M., KETTNER A., "Sediment flux and the Anthropocene", *Philosophical Transactions of the Royal Society A, Mathematical, Physical and Engineering Sciences*, no. 369, pp. 957–975, 2011.

[SYV 14] SYVITSKI J.P.M., KETTNER A.J., OVEREEM I. *et al.*, "Anthropocene metamorphosis of the Indus Delta and lower floodplain", *The Anthropocene*, no. 3, pp. 24–35, 2014.

[SZA 15] SZABO S., RENAUD F.G., HOSSAIN M.S. *et al.*, "Sustainable development goals offer new opportunities for tropical delta regions", *Environment: Science and Policy for Sustainable Development*, vol. 57, no. 4, pp. 16–23, 2015.

[TA 05] TA T.K.O., NGUYEN V.L., TATEISHI M. *et al.*, "Holocene delta evolution and depositional models of the Mekong river delta, southern Vietnam", *River Deltas – Concepts, Models, and Examples*, SEPM, Special Publication, no. 83, pp. 453–466, 2005.

[TA 02] TA T.K.O., NGUYEN V.L., TATEISHI M. *et al.*, "Holocene delta evolution and sediment discharge of the Mekong River, southern Vietnam", *Quaternary Science Reviews*, no. 21, pp. 1807–1819, 2002.

[TEL 02] TELLER J.T., LEVERINGTON D.W., MANN J.D., "Freshwater outbursts to the oceans from glacial Lake Agassiz and their role in climate change during the last deglaciation", *Quaternary Science Reviews*, no. 21, pp. 879–887, 2002.

[TEN 12] TENA A., BATALLA R.J., VERICAT D., "Reach-scale suspended sediment balance downstream from dams in a large Mediterranean river", *Hydrological Sciences Journal*, vol. 57, no. 5, pp. 831–849, 2012.

[TEN 14] TENA A., VERICAT D., BATALLA R.J., "Suspended sediment dynamics during flushing flows in a large impounded river (the lower River Ebro)", *Journal of Soils and Sediments*, vol. 14, no. 12, pp. 2057–2069, 2014.

[TES 15] TESSLER Z.D., VÖRÖSMARTY C.J., GROSSBERG M. *et al.*, "Profiling risk and sustainability in coastal deltas of the world", *Science*, vol. 349, no. 6258, pp. 638–643, 2015.

[TRA 10] TRAN N.N., "The Vietnamese Mekong Delta facing the double challenge of the climate change, International Commission on Large Dams", *78th Annual Meeting*, Hanoi, Vietnam, 2010.

[TUO 03] TUONG T.P., KAM S.P., HOANH C.T. *et al.*, "Impact of seawater intrusion control on the environment, land use and household incomes in a coastal area", *Paddy Water Environment*, no. 1, pp. 65–73, 2003.

[UNI 09] UNITED NATIONS, Terminology on Disaster Risk Reduction, Report, UNISDR, available at: http://www.unisdr.org/we/inform/terminology, May, 2009.

[UNI 11] UNION OF CONCERNED SCIENTISTS, "Ganges-Brahmaputra delta, Bangladesh", Climate Hot Map, available at: http://www.climatehotmap.org/global-warming-locations/ganges-brahmaputra-delta-bangladesh.html, 2011.

[UNI 15] UNITED NATIONS, Sendai Framework for Disaster Risk Reduction and Resilience 2015–2030, UNISDR, Geneva, 2015.

[VAR 97] VARDY S.R., WARNER B.G., ARAVENA R., "Holocene climate effects on the development of a peatland on the Tuktoyaktuk Peninsula, Northwest Territories", *Quaternary Research*, vol. 47, no. 1, pp. 90–104, 1997.

[VEY 10] VEYRET Y., ""Écocatastrophisme" et développement durable", in BRUNEL S., PITTE J.-R. (eds), *Le ciel ne va pas nous tomber sur la tête*, JC. Lattès, Paris, 2010.

[VID 10] VIDAL J., "Outrage at UN decision to exonerate Shell for oil pollution in Niger delta", *The Guardian*, August 22, 2010.

[VRI 96] VRIES I. (DE), SMAAL A.C., NIENHUIS P.H. *et al.*, "Estuarine management strategies and the predictability of ecosystem changes", *Journal of Coastal Conservation*, no. 2, pp. 139–148, 1996.

[WAD 12] WADA Y., BEEK L.P.H. (VAN), SPERNA WEILAND F.C. *et al.*, "Past and future contribution of global groundwater depletion to sea-level rise", *Geophysical Research Letters*, no. 39, article no. L09402, 2012.

[WAS 04] WASSMANN R., HIEN N.X., HOANH C.T. *et al.*, "Sea level rise affecting the Vietnamese Mekong delta: Water elevation in the flood season and implications for rice production", *Climatic Change*, no. 66, pp. 89–10, 2004.

[WES 14] WESENBEECK B.K. (VAN), MURLDER J.P.M., MARCHAND M. *et al.*, "Damming deltas: A practice of the past? Towards nature-based flood defenses", *Estuarine, Coastal and Shelf Science*, no. 140, pp. 1–6, 2014.

[WIK 18a] WIKIPEDIA, "Inondation de 1809 aux Pays-Bas", available at: https://fr.wikipedia.org/wiki/Inondation_de_1809_aux_Pays-Bas, 2018.

[WIK 18b] WIKIPEDIA, "North Sea Flood of 1953", available at: https://en.wikipedia.org/wiki/North_Sea_flood_of_1953, 2018.

[WIL 14] WILD T.B., LOUCKS D.P., "Managing flow, sediment, and hydropower regimes in the Sre Pok, Se San, and Se Kong Rivers of the Mekong basin", *Water Resources Research*, no. 50, pp. 5141–5157, 2014.

[WIL 30] WILLCOCKS W., *Lectures on the Ancient System of Irrigation in Bengal and its Application to Modern Problems*, Calcutta University Press, Calcutta, 1930.

[WIT 57] WITTFOGEL K.A., *Oriental Despotism: A Comparative Study of Total Power*, Yale University Press, New Haven, 1957.

[WOL 00] WOLANSKI E., SPAGNOL S., "Environmental degradation by mud in tropical estuaries", *Regional Environmental Science*, no. 1, pp. 152–162, 2000.

[XIQ 05] XIQING C., ERFENG Z., HONGQIANG M. *et al.*, "Preliminary analysis of human impacts on sediment discharges from the Yangtze, China, into the Sea", *Journal of Coastal Research*, vol. 21, no. 3, pp. 515–521, 2005.

[XU 98] XU J., "Naturally and anthropogenically accelerated sedimentation in the Lower Yellow River, China, over the past 13,000 years", *Geografiska Annaler*, vol. 80 A, no. 1, pp. 67–78, 1998.

[XU 14] XU Y.J., "Rethinking the Mississippi River diversion for effective capture of riverine sediments", in XU Y.J., ALLISON M.A., BENTLEY S.J. *et al.* (eds), *Sediment Dynamics From the Summit to the Sea, Proceedings of a Symposium Held in New Orleans, Louisiana, USA, 11–14 December 2014*, IAHS Publication no. 367, 2014.

[ZAH 13] ZAHARIA L., IOANA-TOROIMAC G., "Romanian Danube River management: Impacts and perspectives", in ARNAUD-FASSETTA G., MASSON E., REYNARD E. (eds), *European Continental Hydrosystems Under Changing Water Policy*, Verlag Dr Friedrich Pfeil, Munich, 2013.

[ZAN 12] ZANINETTI J.-M., COLTEN C.E., "Shrinking New Orleans: Post-Katrina population adjustments", *Urban Geography*, vol. 33, no. 5, pp. 675–699, 2012.

[ZAN 13] ZANINETTI J.-M., "Adaptation urbaine post-catastrophe : la recomposition territoriale de La Nouvelle-Orléans", *Cybergeo: European Journal of Geography, Espace, Société, Territoire*, no. 636, available at: http://cybergeo. revues.org/25805, 2013.

[ZOU 15] ZOU L., KENT J., LAM S.N. *et al.*, "Evaluating land subsidence rates and their implications for land loss in the Lower Mississippi River Basin", *Water*, vol. 8, no. 1, pp. 1–15, 2015.

Index of Place Names

A, B, C

D, E, F

G, H, I

J, K, L

M, N, O

P, Q, R

S, T, U

V, W, X, Y, Z

Index of Common Words

G, H, I, L

M, N, P

R, S, T

U, V, W

Other titles from

iSTE

2019

BOURRIÉ Guilhem
Soils Set
Volume 3 – Soils as a Key Component of the Critical Zone 3: Soils and Water Circulation
Volume 4 – Soils as a Key Component of the Critical Zone 4: Soils and Water Quality

BRAVARD Jean-Paul
Sedimentary Crisis at the Global Scale 1: Large Rivers, from Abundance to Scarcity

KARA Mohamed Hichem, QUIGNARD Jean-Pierre
Fishes in Lagoons and Estuaries in the Mediterranean 1: Diversity, Bioecology and Exploitation
Fishes in Lagoons and Estuaries in the Mediterranean 2: Sedentary Fish
Fishes in Lagoons and Estuaries in the Mediterranean 3A: Migratory Fish

VALENTIN Christian
Soils Set
Volume 5 – Soils as a Key Component of the Critical Zone 5: Degradation and Rehabilitation

LEMANCEAU Philippe, BLOUIN Manuel
Soils Set
Volume 6 – Soils as a Key Component of the Critical Zone 6: Ecology

2018

BERTHELIN Jacques, VALENTIN Christian, MUNCH Jean Charles
Soils Set
Volume 1 – Soils as a Key Component of the Critical Zone 1: Functions and Services

DHÉRISSARD Guillaume
Soils Set
Volume 2 – Soils as a Key Component of the Critical Zone 2: Societal Issues

GUILLOUX Bleuenn
Marine Genetic Resources, R&D and the Law 1: Complex Objects of Use

2017

QUEVAUVILLER Philippe, CIAVOLA Paolo, GARNIER Emmanuel
Management of the Effects of Coastal Storms: Policy, Scientific and Historical Perspectives

2016

BLAIN Stéphane, TAGLIABUE Alessandro
Iron Cycle in Oceans

BRAUDEAU Erik, ASSI Amjad, MOHTAR Rabi
Hydrostructural Pedology

MONACO André, PROUZET Patrick
Seas and Oceans Set
Volume 9 – Tools for Oceanography and Ecosystemic Modeling

QUÉGUINER Bernard
The Biogeochemical Cycle of Silicon in the Ocean

QUEVAUVILLER Philippe
Marine Chemical Monitoring

2015

HERBIN Hervé, DUBUISSON Philippe
Infrared Observation of Earth's Atmosphere

MERCIER-LAURENT Eunika
The Innovation Biosphere

MONACO André, PROUZET Patrick
Seas and Oceans Set
Volume 7 – Marine Ecosystems: Diversity and Functions
Volume 8 – Governance of Seas and Oceans

2014

DE LARMINAT Philippe
Climate Change: Identification and Projections

MONACO André, PROUZET Patrick
Seas and Oceans Set
Volume 1 – Ocean in the Earth System
Volume 2 – The Land-Sea Interactions
Volume 3 – Ecosystem Sustainability and Global Change
Volume 4 – Vulnerability of Coastal Ecosystems and Adaptation
Volume 5 – Development of Marine Resources
Volume 6 – Value and Economy of Marine Resources

MUTTIN Frédéric
Marine Coastal and Water Pollutions: Oil Spill Studies

2013

LÉCUYER Christophe
Water on Earth: Physicochemical and Biological Properties

2012

LE MENN Marc
Instrumentation and Metrology in Oceanography

Printed and bound by CPI Group (UK) Ltd, Croydon, CR0 4YY